The Wisdom of the Genes

THE WISDOM OF THE GENES

NEW PATHWAYS IN EVOLUTION

CHRISTOPHER WILLS

Basic Books, Inc., Publishers

NEW YORK

Grateful acknowledgment is made for permission to reprint:

Excerpt from "Design" copyright © 1936 by Robert Frost and renewed 1964 by Lesley Frost Ballantine. Reprinted from *The Poetry of Robert Frost*, edited by Edward Connery Lathem, by permission of Henry Holt and Company, Inc.

Excerpt from "Evolutionary Hymn" from *Poems* by C. S. Lewis, edited by Walter Hooper, copyright © 1964 by the Executors of the Estate of C. S. Lewis, reprinted by permission of Harcourt Brace Jovanovich, Inc.

Library of Congress Cataloging-in-Publication Data

Wills, Christopher.
   The wisdom of the genes : emerging patterns in evolution
Christopher Wills.
     p. cm.
   Bibliographical references: p. 317
   Includes index.
   ISBN 0–465–09195–4
   1. Evolution. 2. Genetics. I. Title.
QH371.W55   1989
575–dc20   89–42526
                                        CIP

Copyright © 1989 by Christopher Wills
Printed in the United States of America
Designed by Vincent Torre
89 90 91 92 HC 9 8 7 6 5 4 3 2 1

To Liz and Winkie,

two of evolution's finest products

Contents

LIST OF ILLUSTRATIONS   ix
PREFACE   xi

Introduction   3

1   Some Evolutionary Mysteries   11

2   A New Way of Looking at Evolution?   45

3   How Not to Think about Evolution   59

4   The Role of Mutation   102

5   Of Tuxedos and Antibodies   135

6   An Evolutionary Toolbox   159

7   Of Butterflies and Handbags   180

8   The Evolutionary Orient Express   210

9   The Increasing Sophistication of Evolution   239

10 Potential-Realizing and Potential-Altering
    Mutations   273

11 Closing the Circle   288

REFERENCES   317
GLOSSARY   326
INDEX   336

List of Illustrations

1.1 A spectacular Hawaiian fruit fly and its common relative   14

1.2 Portrait of DNA coding for ribosomal RNA   21

1.3 The Duffy-negative allele fights to survive   28

1.4 Skulls of two sabertooths: a case of parallel evolution   41

3.1 Growing old: aging oysters suffer problems   76

3.2 Foraminifera from 150 million years apart: the same old story   80–81

3.3 Lake-dwelling molluscs respond to environmental upheavals   88

3.4 One species changes into another over a half-million-year period   95

5.1 Unequal crossing-over: the benefits and perils of inaccurate genetic recombination   147

5.2 Joining different genes to make a mature antibody   152

6.1 Horse and yeast alcohol dehydrogenase: a comparison   162

6.2 Three-dimensional view of the structure of an enzyme   172

6.3   Genetic Lego: building numerous enzymes
with a small repertory of building blocks   175

7.1   A versatile flounder mimicking a variety of
backgrounds   182

7.2   Butterfly look-alikes: the monarch and the
viceroy   189

7.3   The butterfly *Hypolimnas dubius:* two forms, one
species   194

7.4   The butterfly *Papilio dardanus* and one of the
many butterflies she can mimic   201

8.1   Giant chromosomes of the common fruit fly,
with some of their jumping genes   225

8.2   The body plan of a normal and mutant fruit fly   228

9.1   The ''creationist'' experiment   246

9.2   Snapdragons: plain and spotted   250

9.3   Making genes jump in the somatic cells of a
fruit fly   267

11.1   Typical pelycosaurs: our early ancestors?   293

11.2   An advanced therapsid: almost a mammal   295

11.3   Wild-type silkworm larva and its many-legged
mutant   303

11.4   The aurochs, an ancestor of cattle   311

Preface

I have been teaching evolution to upper-level college students for some twenty years. In the process of digging back through my old dog-eared notes from two decades ago, I find that while a few topics remain the same, the overall emphasis has changed dramatically. Whole new areas of study have appeared. Little was known about the evolution of proteins then. That subject fills many books and monographs now. The use of a data bank of sequences of DNA to follow evolutionary relationships is commonplace now but was unthought of then. Jumping genes, the ubiquitous pieces of DNA that move about the chromosomes, have now been found to play an important role in evolution. Twenty years ago they appeared to be simply an oddity confined to corn plants and were written off by most scientists as some unique peculiarity of corn chromosomes.

In spite of all these advances, there are still huge gaps in our knowledge. We scientists are confident that we have a pretty good bare-bones idea of how evolution works. But it is a long way from our theory to the extremely complex manifestations of life that have to be explained. I will begin the book with this point by introducing a small sampling of some interesting and challenging evolutionary problems. This will be accompanied by a quick sketch of how we currently understand the workings of evolution. The gap will quickly become apparent. A student or a lay person confronted by such a sketch would have, I think, a hard time relating the explanation to the facts that have to be explained. I will then try to close this gap a bit by relating the process of evolution to other processes we are all familiar with. The learning curve that is followed as a technology moves from its first crude applications to a highly advanced form has, as we will see, its parallels in the process of evolution. In subtle and remarkable ways that this book will explore, the process of evolution is becoming easier.

The implications of this way of looking at evolution are large. Once we understand *how* the process is becoming easier, it will actually become easier for us to harness it. For this is indeed what we are about to do. The harnessing of evolution—not just through plant and animal breeding, which we have been doing for millennia, but through the direct manipulation of the genes themselves—is one of the new and breathtaking steps that humankind is already beginning to take.

As a result of what we are learning, it is becoming possible to modify organisms dramatically and give them quite new properties. The first steps have already been taken. Recently it has become technically feasible to mix together cells of early embryos of a sheep and a goat and to plant this new construct into the uterus of a receptive female sheep. Some of the embryos are carried to term and continue to grow into adult *chimaeras.* The monster of Greek myth from which this kind of creature takes its name had a lion's head, a goat's body, and a serpent's tail. The latter-day chimaera does not go quite this far, but it is a mixture of the cells of two very different animals whose last common ancestor lived five million years ago. The adult chimaera has patches of wool and goat hair, and all its internal organs are a mixture of sheep and goat. The poor beast looks at us dolefully from the pages of the newspaper, a victim of the ultimate identity crisis.

Even more distantly related creatures can be mixed and matched. Just down the hall from my laboratory at the University of California, San Diego, a group of molecular biologists headed by Stephen H. Howell recently succeeded in putting the gene for the enzyme luciferase, taken from a firefly, into tobacco plants. When these plants are supplied with the substrates for the enzyme, luciferin and ATP, they glow. The light is feeble, but easily perceptible to the dark-adapted eye. Much hilarity immediately resulted from irreverent suggestions for uses for the plants. (One possibility is to relandscape Central Park with them, lighting up the muggers on their nefarious rounds.) But in fact, this unlikely wedding of a plant and an insect is already starting to pay off scientifically. The development of different parts of a plant can be followed by using light production as a marker, and the enzyme can actually be inserted into different regions of the cell so that internal cellular development can be followed in detail.

Other scientists on this campus and elsewhere are working on the task of inserting copies of a functioning gene into a patient with a disease caused by a lack of that same gene function. The basic science has been done, and already this "gene therapy" has been carried out with limited success in mice. The remaining problems, though numerous, seem to be merely technical.

As with any new technology, the first attempts are clumsy. But we are playing with the very stuff of life itself. We cannot, in view of the many unknown consequences of producing new life forms, afford to go on being clumsy indefinitely.

At the end of the book I will suggest some ways in which we might improve our chances for success in taking charge of the evolutionary process. If indeed the genes have grown wiser with time, then so must we. It is inevitable that we will use these new tools. But it would be extremely incautious of us to make this technological leap without thinking through the consequences, the way we did at the start of the atomic age.

This book has grown out of conversations with many of my fellow scientists, who will probably find much to disagree with here. The book itself has been read in its entirety by my colleagues Ted Case, Russell Doolittle, Mark Martin, and Trevor Price, who made many valuable suggestions and encouraging noises. The usual disclaimer that the author makes at this point is that his friends made many suggestions, not all of which he chose to follow. But their comments were so valuable that I really did try to incorporate them all. My daughter, Winkie, a merciless critic of parts of the book, never lost the chance to stick it to her old dad if she did not like a turn of phrase or did not understand something. Charles Cavaliere and Richard Liebmann-Smith expertly saw the book through production and caught a number of gaffes at the last minute. I would also like to thank my agent, Gerard McCauley, whose patience and belief in the book resulted in its finding a secure berth.

Some suggested readings, grouped according to chapter, are found at the end of the book. I have further separated the readings into general books and articles, and a selection of the more specialized research papers on which the recent work discussed here is based. I have also added a glossary at the end of the book that may help guide the reader through the unavoidable thicket of biological terms.

The Wisdom of the Genes

Introduction

Teleology is a lady without whom no biologist can live. Yet he is ashamed to show himself with her.

—E. VON BRUECKE

On the night of Bastille Day, 1918, the American physiologist Walter B. Cannon was catching a few hours of exhausted sleep on the line at Châlons-sur-Marne, just south of Reims. Cannon's skills as a doctor had been tested to the utmost over the previous few months, as the American Expeditionary Force began the immense task of breaking the deadlock reached by its war-weary Allies and pushing the Germans back from positions they had occupied for almost four years. Like the men on the line, he had been subject to stress almost beyond human endurance.

Just after midnight, and with appalling suddenness, the last desperate German offensive of the war began. The bombardment started instantaneously, in Cannon's words, "like Niagara behind the falls, like thousands of huge motor trucks speeding over cobblestones, like freight trains passing each other in a tunnel." Within minutes of being jerked awake he found himself trying to save a flood of hideously wounded men, in shock both from their injuries and from the din of the attack.

Before the day was out, he would survive repeated shellings of the field hospital. But the line held, and at last the war reached a turning point.

During all this time, Cannon's cool powers of observation did not desert him. Almost all the wounded men he saw suffered from extreme thirst, made worse by an inability to swallow liquids. He had earlier noted that these symptoms were correlated with a severe acidosis of the blood, and this had led him to a crude but effective treatment—the direct injection of a strong solution of sodium bicarbonate. This drastic treatment worked in many cases, though only for a short time. Problems with blood group incompatibility and the primitive technology available made transfusions, which could treat the problem at a more fundamental level, difficult and chancy.

Normally, Cannon knew, the body is able to survive even quite extreme physiological stress. But trench warfare produced stresses that were beyond the capability of the body to withstand without some external help. Giving that help, even in the crude form of injecting a strong solution of buffer, allowed the natural balance of forces within the body to reassert itself. As he observed men fighting to survive after being badly wounded, the thing that impressed him more than any other was the ability of the body to reestablish the conditions necessary for survival.

After the war, Cannon returned to a life of active research and teaching at Harvard Medical School. In 1932 he summarized his years of physiological research and the observations of his wartime experiences in a little book called *The Wisdom of the Body*. The title was actually borrowed from a lecture that had been given by an English physiologist, E. H. Starling, some years earlier. Cannon's book explored the many ways in which the body, despite enormous external stresses, can keep its internal conditions surprisingly constant. Blood loss can be compensated for through withdrawal of liquid from the lymphatic system and the tissues. Temperature can be maintained within narrow limits through di-

version of blood flow to different areas. Emotions such as
fear and rage produce corresponding physiological effects that
can enhance survival. He lumped all these compensating
mechanisms together under the term *homeostasis,* a word that
has entered the language.

I have adapted his title for this book because the message
I would like to convey is a very similar one. Cannon was
concerned with mechanisms leading to the survival of the
body. This book is concerned with the mechanisms that lead
to the survival of the genes. The wisdom of the body is the
result of the wisdom of the genes, and that in turn is the
result of billions of years of refinement through evolution.

[margin note: now 50 million]

Estimates of the number of species of animals, plants, and
bacteria on the planet at the present time vary, but the number has been put as high as ten million. Each species, no
matter how simple or complex, has a history of three and a
half billion years, extending back to the first organisms that
could truly be called alive. And each has a history so fraught
with perils that the odds against survival are almost vanishingly small. Indeed, the overwhelming majority of species
have become extinct, through sheer chance, the inability to
adapt, or both. The ones that are left are, for the most part,
a superbly fit set of survivors.

In the last three and a half billion years, there must have
been many times when our own ancestors were subjected to
environmental shocks as severe as those suffered by Cannon's soldiers. The history of each of us is a history of millions of such hairsbreadth escapes, and each of these has had
a part in shaping our genes and making them better at their
job.

Hence the wisdom of the genes. But we must be careful
with the implications of this phrase. The word *wisdom* comes
with a freight of meanings because it describes such a pe-
culiarly human attribute. Only rarely do we extend the idea
of wisdom to the nonhuman world, perhaps in talking about
a wise old dog or more vaguely about the wisdom of the

universe. It is dangerously anthropocentric to talk about the genes being wise, because it appears to provide them with abilities, perhaps including the ability to pick out and work toward a goal. But Cannon and I would have been able to avoid the teleological implications of the word *wisdom* only by giving our books more insipid titles. This would in turn cause the reader's eye to glide effortlessly past them as they sit gathering dust on the shelf.

So it must be emphasized at the outset that the genes are only wise in the sense that the body is. Just as the body is capable of adapting to many different short-term environmental fluctuations, so the genes are capable of evolving as the environment changes gradually over more extended periods of time.

The theme of this book is that there is an accumulated wisdom of the genes that actually makes them better at evolving (and sometimes makes them better at not evolving) than were the genes of our distant ancestors. This does not mean that the genes are becoming smarter with time; rather, this apparent accumulation of wisdom is due to, and is indeed an inevitable result of, the forces of evolution that scientists are already familiar with. There is nothing mystical or anthropocentric about it.

Simply put, evolution is getting easier with time by means of a process that we might call *evolutionary facilitation*. During the more than 3 billion years that life has been on the earth, it has evolved into incredible complexity. In the process the genes of many organisms have become altered, grouped, and rearranged into patterns that actually facilitate further evolution. Some organisms—and some genes within organisms—have become experts at evolving, and at the same time others have become experts at not evolving. This growing expertise requires not only new levels of gene organization, but also new kinds of mutations, or genetic changes. These are supplied by new kinds of mutation-causing agents, which have themselves evolved to become more complex and subtle in their interactions with the genes that they affect.

Mutations fuel the evolutionary process. The idea of evolutionary facilitation implies that mutagenic agents, the factors that cause mutations, have over time become more likely to produce the kinds of phenotypic changes that are advantageous. This can happen for two reasons. First, many mutational changes are now known to be produced by factors in the cells themselves. These include parasites that actually live in our DNA. These parasites and other factors are more likely to survive if they produce advantageous rather than disadvantageous changes in their hosts' genes. Second, the genes with which these mutational agents interact have become shaped in the course of evolution by this very interaction, with the result that useful mutations are more likely to occur. In short, the interaction between the mutation-producing factors and the genome is itself subject to evolution—an interaction that is a virtually unexplored area of evolutionary biology.

It will become apparent as the story unfolds that tests of the hypothesis that evolving can become easier with time have already been made—recently and notably by John Cairns and his co-workers at Harvard and by Barry Hall at the University of Connecticut. And various aspects of this idea have been suggested by a number of evolutionary thinkers—John H. Campbell, Stephen Jay Gould, Walter Gilbert, and others—though none has put it as baldly as I have just done.

To understand this idea, we must explore some of the most exciting areas in modern biology. The old observations of nineteenth-century biologists such as Charles Darwin are daily confirmed and extended by new techniques undreamed of only a few years ago. Because of these new techniques, information is being accumulated much more quickly than it can be presented to the public, and indeed the scientists themselves are having a tough time assimilating it. For the first time, it is possible to examine in detail the ways that genes have evolved. We can even use these new techniques to peer billions of years into the past—not all the way to the beginning of life, at least not yet, but our vision is growing clearer with each passing year.

It will be necessary first to take a quick glance at some of the evidence that has accumulated since the time of Darwin about how evolution works. I will try to make it as painless and interesting as possible. There is in any case no way a book this size could cover it all. A friend of mine has been working on such an encyclopedic evolution textbook for years, and I wish him the best of luck: he is in the position of Tristram Shandy, who took a year to write in his diary the events of a single day.

Because of the great mass of material, we must of necessity be selective in the topics chosen. This will allow us to delve into those areas that most clearly display the putative wisdom of the genes. I will show that this wisdom consists both of the ways that genes have become organized in the course of evolution and the ways in which the factors that change the genes have actually become better at their task.

There is, as you will discover, a rather pleasant intellectual fallout associated with this new way of thinking about evolution. In effect, it makes the whole process of evolution itself more understandable.

Scientists and educators have tended to shy away from admitting that it is necessary to convince people that evolution has occurred. It is presented to the public as a *fait accompli.* Evolution is simply one of the foundations of scientific thought, like the new physics, that have made the world more interesting and comprehensible. Yet, we know from the daily news that evolution is different from the new physics in the way that it intersects with the public consciousness.

In a peaceful world atoms are split somewhere else, safely (we hope) sequestered in huge domes of reinforced concrete. Properly harnessed, they provide us with abundant energy, and only occasionally do they run rampant. We try not to think about the possibility of nuclear doom. And most people are blissfully unaware that natural and artificial radioactive elements are present in their own bodies in small amounts and are continuously decaying.

But evolution is different. Evolutionists purport to explain where we came from and how we developed into the complex organisms that we are. Physicists, by and large, do not. So, the study of evolution trespasses on the bailiwick of religion. And it has something else in common with religion. It is almost as hard for scientists to demonstrate evolution to the lay public as it would be for churchmen to prove transubstantiation or the virginity of Mary. The process of evolution, though it pervades our lives, is subtle and hard to detect.

There is no need to convince the public that atoms can be split and that the relation between matter and energy is just what Einstein predicted. Although they may not know the details, they can certainly see the results. In contrast, unless we are plant or animal breeders, amateur or professional naturalists, or scientists directly concerned with the process of evolution in the laboratory, we do not see evolution at work.

In fact, a very vocal minority of fundamentalist Christians, known as *creationists,* has capitalized on this public uncertainty about how and whether evolution takes place. Their original demand was that students should be denied information about evolution. After many setbacks in the courts they now grudgingly admit that it should be provided, but only when paired with information about their own particular subset of fundamentalist beliefs concerning the origin of living things. Their hope appears to be that the current vagueness on the part of young people about *how* evolution works (and even *if* it does) might somehow be channeled into a kind of mass conversion of the uncommitted to a fundamentalist faith.

I rather doubt that their hope will come to pass. But they have succeeded in doing a great deal of damage. Partly as a result of their influence, generations of American school-children have been exposed to timid textbooks and uncertain teachers. This has had the indirect result of casting doubt on the likelihood of the evolutionary process. When it is taught in a superficial way, evolution becomes just as much of a fairy story as the creation myths of the Book of Genesis.

The process of evolution is complicated. The results of this process appear at first sight to be rather unlikely. Indeed, looked at superficially, there is surely nothing more unlikely than the evolution of the human eye—as the creationists never tire of telling us. But the evolution of such complex structures is not as improbable as the creationists would have us believe. In his recent book *The Blind Watchmaker,* Richard Dawkins has emphasized that even very complex structures and behaviors can be built up by selection through small incremental steps. I would like to go further and suggest that some kinds of evolution can actually become easier with time. This is because each step in the process draws on the accumulated wisdom of the genes.

1

Some Evolutionary Mysteries

The closer one looks at these performances of matter in living organisms, the more impressive the show becomes. The meanest living cell becomes a magic puzzle box full of elaborate and changing molecules, and far outstrips all chemical laboratories of man in the skill of organic synthesis performed with expedition and good judgment of balance . . . [A]ny living cell carries with it the experiences of a billion years of experimentation by its ancestors. You cannot expect to explain so wise an old bird in a few simple words.

—Max Delbrück, *A Physicist Looks at Biology*

AN ADAPTIVE EXPLOSION

We halted in a leafy glade in an old part of the rain forest. Here the 'ohia trees had matured, and the whole area was rather parklike. Lava flows from nearby Kilauea had not come this way for many years. The warm Hawaiian air was very still, and the shade under the trees was deep.

We were here to look at one of the most remarkable examples of evolution in action to be found on the planet. My companion invited me to drop on my hands and knees. There, scurrying around in the leaf litter, were tiny dark insects. Some had prominent white markings on their upper bodies; others were almost unnoticeable against the brown leaves.

When I looked closely, I saw that they were flies. But what

little flying they did was in short hops. Some of them spent all their time in the leaf litter, while others crawled over the leaves and branches of the nearby bushes. Their lives seemed very busy and complex, and it was certainly unusual to see flies living in places where we would normally expect to find beetles or pill bugs.

These were, of course, not ordinary flies. They had taken up their life among the fallen leaves because there were very few other animals to compete with them for this rich and varied ecological niche.

The Hawaiian Islands, specks in the middle of the Pacific, are the most isolated pieces of land on the planet. Islands in this chain have been in existence for many millions of years, although the current group of islands is only a few million years old. The big island of Hawaii, on which we stood, is the youngest. This squalling infant of an island (Kilauea has just finished erupting again as I write this) is a mere seven hundred thousand years old.

The islands have formed, one after the other, as the central Pacific tectonic plate slides across a "hot spot" in the molten mantle beneath. Mauna Loa and Mauna Kea, the two mighty volcanoes on the big island, are almost fourteen thousand feet tall. But as one travels west the islands get smaller and lower. Erosion is no longer counterbalanced by volcanic activity. Finally, beyond Niihau, some four hundred miles west of where we were standing, the islands have been reduced by erosion to stumps that in some cases do not even reach the surface of the ocean as reefs. The stumps become older and older as the chain surfaces briefly some fifteen hundred miles away, at Midway, then veers north toward the Aleutians.

During all these millions of years, only about three hundred different species of insect have made the long, accidental journey to the islands, borne on the wind or carried by birds. Other animals and plants have arrived the same way. Still others were carried from Asia by unusual storms that also deposited fertile dust on the barren volcanic ash of the islands. Some may have rafted on logs from the Pacific Northwest—logs from the Columbia River still occasionally wash up on Hawaiian beaches. Most of these scattered arrivals have since evolved into forms unique to the islands, endemic species that are found nowhere else in the world.

The flies of Hawaii belong to the genus *Drosophila*. They are related to the geneticists' familiar fruit flies, about which we will have a lot to say in this book. There are many species in the genus, probably about twenty-five hundred worldwide. Remarkably, a full third of these are found on the Hawaiian Islands, and they are all closely related to each other. It is actually possible that all these eight hundred species are descended from a single inseminated female, carried to the islands an unknown number of millions of years ago.* Because females of these flies store sperm in a special sac, they need only mate once and are able to lay fertile eggs for the rest of their reproductive life. Thus, it would take a minimum of one fertilized female, landing in this Garden of Eden, to be the ancestress of all the species that we now see.

The results of this remote immigration are by now extremely varied in appearance. They live on many different kinds of native plants. Some are enormous by fruit fly standards, up to an inch long. Some have strange broad heads with eyes positioned far apart like those of hammerhead sharks. One large group exhibits a variety of striking dark patterns on their wings, and as a consequence are called the picture-winged flies. (Figure 1.1 shows one of the more remarkable Hawaiian Drosophila, with the common fruit fly, *Drosophila melanogaster*, as a comparison.)

Their behavior also varies. In some species, the males con-

* Recent molecular evidence suggests that there may have been more than one introduction, the earliest as much as 60 million years ago, when the present-day Hawaiian Islands had not even appeared. The story is not yet complete, however, and the Hawaiian Drosophila appear to be more closely related to each other than to any other Drosophila group.

Figure 1.1 One of the larger and more spectacular Hawaiian Drosophila, *D. cyrtoloma*, with the common fruit fly of geneticists, *Drosophila melanogaster*, as a comparison. Both flies are magnified 10×. (From figure 5 of H. L. Carson, D. E. Hardy, H. T. Spieth, and W. S. Stone. The evolutionary biology of the Hawaiian Drosophilidae, in *Essays in evolution and genetics in honor of Theodosius Dobzhansky*, ed. M. K. Hecht and W. C. Steere [New York: Appleton-Century-Crofts, 1970], pp. 437–544.)

gregate at specific sites called leks, where they put on complex displays to attract the females. In one of the more spectacular of these displays, the males of the species *Drosophila hamifera* rush at each other like knights in a medieval tourney. The goal is to knock each other off the branch they are occupying. In a number of other species the males rear up on their hind four legs and engage in wrestling matches. They sometimes strain at each other for five minutes before one is overturned or flees.

In spite of all this variation in appearance and behavior, it is easy to determine that these species are closely related to each other. Their chromosomes are so similar that only an expert can tell them apart, and sometimes there appear to be no chromosomal differences at all between species. Further, even though flies of the various species may appear to be very different, they can be shown to share proteins to a considerable degree. Indeed, were chromosomal or protein comparisons the only criteria for classifying species within each major group, taxonomists would put them in the same species.

My companion in the forest glade, Hampton Carson of the University of Hawaii, has spent much of his life studying these flies in all their immense variety. That evening we talked at length about how such explosive evolution could have occurred. It was obvious that these flies are genetically malleable—perfect subjects for evolution. From that remote common ancestor or ancestors they have diverged to fill many different niches, a process called adaptive radiation. But would any other insects, tossed into the same situation, radiate in so remarkable a way? Are there some insects that are better at evolving than others? And if so, why? How can organisms become good at evolving?

We shied away from the implications of this line of reasoning, because they led us into areas where good evolutionists tend not to go. This is because such thoughts smack of the evolutionists' equivalent of original sin: Lamarckism.

But in the intervening years I have returned to these questions again and again. This book is an attempt to answer them, and to do so without the taint of original sin.

THE NUTS AND BOLTS OF EVOLUTION

When Darwin published *The Origin of Species* in 1859, he introduced the term *natural selection*. But nowhere did he use the word *evolution*, although it is assumed by the public that Darwin alone is responsible for the idea of evolution. In fact, of course, no scientist is solely responsible for any scientific idea. Each new idea is in part a synthesis of many older observations and hypotheses, and it is always possible to find hints of any new insight in the work that came before.

The idea of natural selection, Darwin's primary contribution, had certainly been hinted at, sometimes quite broadly, by others. Oddly, the ones to come closest were the scientists and philosophers concerned with the human condition. For example, in 1798 a clergyman named T. R. Malthus pointed out with some relish that unpleasant things would inevitably happen to any human population that was allowed to outstrip its resources. Though he later modified his view slightly, his grim prognosis of the inevitable fate of the poor and overcrowded had tremendous impact. Darwin, looking at Malthus's book from the point of view of a biologist, found that it greatly clarified his thinking about the dynamics and the fate of natural populations in general. If the human population were prevented by natural pressures from outstripping its resources, then surely populations of other organisms would be similarly constrained. It was not far from that to the idea of natural selection.

Indeed, the idea hung half-formed in the air of the mid-nineteenth century, ready to be picked up by astute minds.

One of these belonged to the philosopher Herbert Spencer, a kind of scientific groupie of his day. Brilliant but erratic, he hung around the fringes of biological science and used the half-digested information he obtained to throw out a veritable Catherine wheel of ideas. Most of them, alas, were only remotely founded on fact. But his contacts with the early Victorian scientific community were excellent, and he became great friends with the man who was soon to become Darwin's chief defender, Thomas Henry Huxley.

Spencer's ideas about what is now known as social Darwinism—the role of natural selection in human affairs—were already well formed before Darwin's book appeared. In 1852 (well before *The Origin of Species*) Spencer read a paper before the British Association in which, among much florid argument and unsupported theorizing, there are bits that sound remarkably Darwinian:

> From the beginning, pressure of population has been the proximate cause of progress. . . . For those prematurely carried off must, in the average of cases, be those in whom the power of self preservation is the least, it unavoidably follows, that those left behind to continue the race, are those in whom the power of self-preservation is the greatest—are the select of their generation.

Not bad, except for the bit about progress. Many other evolutionary thinkers, including Darwin, were not above confusing the idea of evolution with the idea of progress. But Darwin's strength lay in his backing up the simple idea that the survival of the fittest would produce evolutionary change with an avalanche of facts, something Spencer could not do. Darwin also pursued the implications of natural selection into areas that none of his contemporaries had thought about— selection for mating success and even selection for the expression of emotions.

Darwin understood the implications of natural selection brilliantly. But he did not know—and knew that he did not know—what was being selected. He did not know about the genes. It was not until the beginning of the twentieth century that an understanding of the mechanics of heredity led to the great synthesis of genetics and evolution. Darwin's ideas were extended and made into a really satisfying theory only when they were combined with the insights of the Moravian monk Gregor Mendel. This happened years after the death of both Darwin and Mendel through the work of a host of geneticists who put Mendel's principles to everyday use.

Mendel, whose paper was published in 1865 and ignored for the next thirty-five years, made some simple generalizations about how genetic information was inherited. These were based on careful experiments with pea plants. Mendel hoped that these generalizations might hold up with other organisms, but he found to his disappointment that they did not seem to work with bees or with another species of plant that he tried. Many of his notes and journals were destroyed after his death, so we will never know the extent of his disappointment or how convinced he was that his laws really were generally applicable.

We now know that Mendel's generalizations are indeed generally applicable, and over a much wider range of organisms than he even knew existed. The apparent exceptions that bedeviled him and other early workers in genetics have now all been explained.

Generations of undergraduates have wrestled with Mendel's laws. They sound a bit quaint as he formulated them, and since that time a great infrastructure of genetic information has been built up around them. But the essence of his laws, like the essence of Darwinism, is simple: (1) genes that are not lost are normally passed unchanged from one generation to the next; (2) they are still there even though their effects in the next generation may be disguised by other genes that control or influence the same trait; and (3) given the right circumstances, they will still yield the same effect in an organism's descendants.

Genes do occasionally change, by mutation or by recombination. During the latter process, they can exchange parts with similar genes. But it is the continuity of genes from one generation to the next that supplies the genetic underpinnings of Darwin's theory. Darwin did not realize it, but it is the genes that are subject to natural selection.

The idea that discrete bits of information were passed from one generation to the next was missing from the widely assumed theories of inheritance that were current in the nineteenth century. Darwin realized this and tried to invent such bits, which he called *gemmules*. He imagined that they were produced by all parts of the body and then migrated to the gonads. Selection for more or fewer of a particular type of gemmule might, he thought, bring about the kind of long-term genetic changes that his theory of natural selection demanded. But his theory was damaged by his cousin Francis Galton, who transfused blood between white and black rabbits, with no effect on their progeny. And it was finally destroyed by the German cytologist August Weismann, who chopped the tails off generations of mice with no effect on the tails of subsequent generations. The genes, as Weismann realized even before the rediscovery of Mendel's work, are safely sequestered inside the nucleus of the cell and out of reach of ordinary environmental effects.

Darwin's first crude attempt to link genetics and evolution had failed, but by the 1930s much more was known about genetics. As a result geneticists and evolutionists began, in a state of high excitement, to collaborate in order to bring their two fields together. A dazzling roster of scientists from many countries was converging on the same set of ideas—a new synthesis, called *neo-Darwinism* by Julian Huxley. (A distinguished evolutionist, Huxley was Thomas Henry's grandson and brother of Aldous, and was later to become the first director general of UNESCO.) This neo-Darwinian synthesis can be made quite complex, but a basic, stripped-down or nuts-and-bolts view goes something like the following.

You have a mind, an existence, perhaps a soul (although we will not deal with this last point in this book). But so far as evolution is concerned, what matters are the genes that you pass on to the next generation, or the genes that you aid in their passage. You might think that if you have no children you will have no effect on the evolutionary process, but this is not true. A childless person may help others to have children, or—like the presumably childless Hitler—prevent many others from having them.

But the genes, of course, are the most important thing. And now we can see what they look like. Figure 1.2 shows a famous picture in the annals of molecular biology, the first photograph of genes in action. It was taken by the brilliant electron microscopist Oscar Miller in 1969, and it shows how the DNA of the genes make RNA. This RNA in turn can make protein, so that the usual flow of genetic information in the cell is from DNA to RNA to protein.

Special techniques allow us to see the long strands of DNA, in this case from a chromosome of a large cell of an amphibian; the cell is on its way to becoming an egg. Such cells carry many copies of genes that are responsible for making the millions of ribosomes that in turn manufacture proteins. All the genes you see in the picture are the same.

The DNA strands, each carrying a series of these genes, meander across the picture. The genes themselves are made visible because many molecules of an RNA-making enzyme are moving along their length. You cannot see the enzymes, but you can see the RNA molecules that they are making, forming a kind of bush of ever-lengthening strands as the procession of enzymes moves down the length of the gene.

As each enzyme making the RNA reaches the end of the gene, both it and the RNA drop off. The enzyme then attaches to the beginning of one of the genes and starts to make a new RNA molecule. You can see that each gene is making many RNA molecules at once. Completed pieces of RNA will

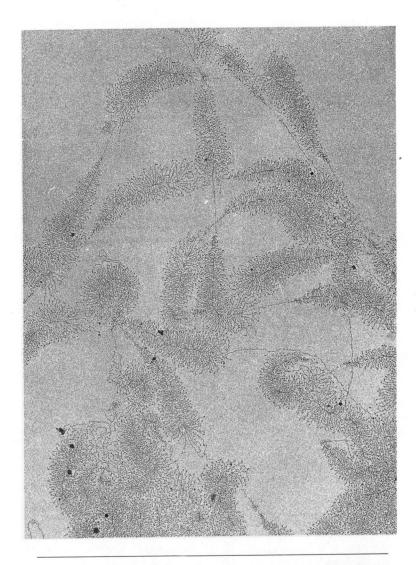

Figure 1.2 Electron micrograph of DNA strands coding for genes for ribosomal RNA. The RNA is being made directly from these strands by many enzyme molecules that move along the DNA, and as the RNA molecules being made from each gene lengthen, they form a kind of bush. Completed RNA strands drop off the ends of the genes, while the enzyme molecules go back to the beginning and start over. The picture is magnified 27,500 times. (From figure 3 of O. L. Miller, Jr. and B. L. Beattie, 1969. Portrait of a gene, *J. Cell Physiol.* 74 Supp. 1 [1969]: 225–32.) Copyright © 1969 by The Wistar Press.

float away, to be incorporated into the growing army of ribosomes in this huge, metabolically active egg cell.

Similar things are going on in our own cells. Most of the cells in our body have forty-six chromosomes in the form of long DNA strands; twenty-three of these are derived from our mothers, and twenty-three from our fathers. Each chromosome consists of many genes strung together. Because we receive one set of chromosomes from each of our parents, our cells contain two copies of each gene. This redundancy of information has interesting consequences, as we will see.

The amount of information carried by these forty-six chromosomes is immense. The DNA is made up of subunits called nucleotides, and there are about six billion of these in the complete human chromosome set, an average of 130 million for each chromosome. All of this DNA would extend about three yards were it to be stretched out fully. It is in fact squeezed into the nucleus of the cell, which is about two millionths of the volume of a pinhead.

The nucleotides are like bits of information in a very long computer program, chopped up into subprograms in the form of genes. But the program is far larger and the density of information far greater than in even our best computers. This becomes apparent when we compare the nucleus of a cell to a computer chip, our most sophisticated electronic device. If we ignore their clumsy electrical connectors to the outside world and the thick substrate on which they rest, the best current computer memory chips hold a million bits of information in a volume one millimeter on a side and ten-millionths of a millimeter thick. This may seem very small, but it is still twenty thousand times the volume of a human nucleus. If the nucleus were similarly inefficient, it could hold only fifty bits of information. But in fact, it holds 240 million times as much as that.*

---

* Because there are four nucleotides in the DNA, each could be represented by two bits—say 00 for A, 01 for T, 10 for G and 11 for C. The fifty

A string of one or two thousand of these nucleotides contains enough information to specify the synthesis of an average protein. If all genes made proteins, there would be enough genes on each chromosome to specify a hundred thousand or so, yielding a total of about two million.

But this would be far more proteins than actually exist in any organism. Even we, the most complicated organisms we know of, are apparently made up of no more than about sixty thousand different kinds of proteins.

Since we produce a relatively small number of proteins compared with the number we could make, it is apparent that genes for proteins account for only a small percentage of the DNA. Much of the rest seems to be involved in other functions, such as regulating the levels of proteins that are made and determining the course of development of individual cells or groups of cells in the body. And indeed, a lot of the DNA seems to have no obvious function at all. All these stretches of DNA, whether we understand what they do or not, can be considered genes in the broadest sense, since careful copies are made by the cell and passed on to the next generation.

The genes make up the organism's *genotype*, while its appearance, function, and behavior make up its *phenotype*. Genes and environment work together to shape the phenotype. For some characters, the genetic contribution predominates, and for others the environment plays the larger role.

The genes we happen to possess are collectively called our *genome*, and the genes that are found in the whole human species make up the human *gene pool.* Just as we borrow a car from a motor pool, we are the brief custodians of a collection of genes from the human gene pool. The analogy

---

bits of information that a nucleus-sized computer chip could hold therefore represent the same information that could be coded by twenty-five nucleotides. Since there are six billion nucleotides in the nucleus, they pack in six billion divided by twenty-five, or 240 million times as much information as the computer chip.

quickly breaks down, however. We are supposed to return the car intact, but we do not pass on our genomes intact to the next generation. Some organisms do, like many flowering plants, but these are the ones that do not reproduce sexually but instead produce clones or copies of themselves.

Because we reproduce sexually, we do not pass on all our genes; rather we give only half of our genes to each offspring, with the rest coming from the other parent. And the one copy of each of our genes that we pass on to our offspring can be either the one we received from our mother or the one we received from our father. As a result, we pass on an overlapping but partially different set of genes to each offspring, which is why our children are all so different from each other (identical twins, triplets, and so on always excepted).

Thus, even if you have children, some of the genes that you carry are sure to be lost. Since the likelihood that a gene is not passed on to your first child is one-half, then the likelihood that it will not be passed on to either your first or your second is one-half times one-half, or one-quarter. Even if you have ten children, which I do not recommend, the chance that a gene will be lost is still one in 1,024 (one-half raised to the tenth power).

If all the genes specifying a particular protein or a particular function in the gene pool were identical, this would make no difference. But often there are two or more forms of a given gene present in the gene pool. Indeed, we now know that most of the genes in the human gene pool come in two or more identifiably different forms called *alleles*. You are already familiar with such alleles. People with blood type AB are *heterozygous* for two different alleles found commonly in the population. Others, by the luck of the draw, may be *homozygous* for type A or for type B. At each genetic locus, we receive from our parents either two copies of one allele or one copy each of two alleles by chance.

People with the normal number of chromosomes cannot of course have more than two alleles at any genetic locus, but there may be many more than two alleles present in the gene pool. In the gene pool of clover, for example, there are more than two hundred different alleles of a gene that controls the growth of the pollen tube through the style of the flower. But, again, no individual plant can carry more than two of these.

RIPPLES IN THE GENE POOL

This great variety of alleles provides the genetic variability in populations, and it is this variability in turn on which natural selection and chance act, increasing the numbers of some alleles and decreasing the numbers of others.

Why are there so many alleles, and where do they all come from? The forces involved were worked out in the 1930s. This was, indeed, the grandest achievement of the neo-Darwinian synthesis. Ways, both physical and mathematical, were found to probe the depths of the gene pool.

A successful new mutation in the gene pool is rather like a stone dropped into a pool of water. The larger the pool, the longer the ripples take to reach the opposite bank. In order for any permanent change in the gene pool to take place, a new mutation must eventually spread through a population and replace the old allele. This may take a long time, and many things may happen. A new mutation, even a highly favorable one, can be lost. Or it might be halted partway through the spreading process. This has happened, for example, to mutations that confer resistance to malaria.

Four major types of malaria afflict the human population. In Africa, the two main types are vivax, or benign tertian malaria, and falciparum, or malignant tertian malaria. In spite of its name, benign malaria can be quite severe, with fevers of up to 105 degrees every other day. These fevers are triggered by waves of parasites released from the red blood cells. But unless the victim is suffering from other life-threatening diseases at the same time, it is usually not fatal. In malignant malaria, however, the parasites are not confined to the bloodstream. They can affect the brain, leading to coma and death. If they escape to the kidneys, they can cause blackwater fever, a late and usually fatal stage of the disease. Placental infections by the parasites often lead to spontaneous abortion.

Benign malaria is debilitating, but malignant malaria kills more than a million people every year, many of them children. And it is on the march again, because of cutbacks in spraying programs and the appearance of strains of parasite resistant to antimalarial drugs.

It is possible to become resistant to malignant malaria, but only after repeated, life-threatening exposure to it. But some people, particularly blacks from the west coast of Africa, are extremely resistant to the benign form of malaria, even if they have never been exposed to it. This resistance is so widespread that the vivax malaria parasite is actually quite rare in tropical Africa. It is much more common in regions farther from the equator, the Mediterranean basin and southern Africa. It is as if the disease itself has been driven out of the tropics by the resistance of the population.

At the beginning of the 1970s it was shown that the native inhabitants of West Africa did not develop vivax malaria even when they were deliberately injected with infected blood. And, in a long-continued series of experiments, Louis Miller and his associates at the National Institutes of Health have obtained strong evidence that this is because they lack a protein on the surface of their red blood cells that is essential to infection. The protein is associated with a carbohydrate that stimulates the formation of antibodies. Together they are known as the Duffy antigen, and people without the protein-carbohydrate complex are Duffy-negative.

The vivax malaria parasite attaches to this protein and uses it to slip inside the red blood cell. But there is nothing on the surface of cells of African blacks for the parasite to grasp, so that this simple genetic change confers complete immunity to the disease. And it does so without harming the carrier of the Duffy-negative gene, who apparently does not need this protein at all to live a normal, healthy life. (What is the protein really for, then? Nobody has any idea.)

By contrast, the falciparum parasite that produces malignant malaria enters the cell in quite a different way. African populations are as susceptible to it as are any others.

Vivax malaria must have been a very widespread disease in tropical Africa thousands of human generations ago. At some point, and quite by chance, a Duffy-negative mutation arose in an inhabitant of the region. Copies of this gene were passed from parent to offspring, perhaps across many generations, until somebody inherited the mutation from both parents. We now know enough to be able to reconstruct the course of this interaction of the human gene pool with mutation and disease (see figure 1.3).

The period leading up to this event was a dangerous time for the Duffy-negative mutation, for there were many points at which all copies of it might have been lost before it became quite by chance common enough to appear as a homozygote. But when such a homozygote for the Duffy mutation did appear, the carrier was completely resistant to the ravages of vivax malaria, and as a consequence was more likely to survive and have children. Slowly at first, and then more and more rapidly as the allele gained in frequency, homozygotes resistant to the disease appeared in the population.

Eventually, more than 99 percent of West African blacks were homozygous for the Duffy-negative allele. But the allele spread no further because the African population remained genetically quite isolated until very recently. Further, vivax malaria was not common enough elsewhere to give the Duffy-negative allele the advantage it needed to spread.

The story did not end there, however. The evolutionary process never stands still. Because of the spread of Duffy-negative, vivax malaria became less prevalent in central Af-

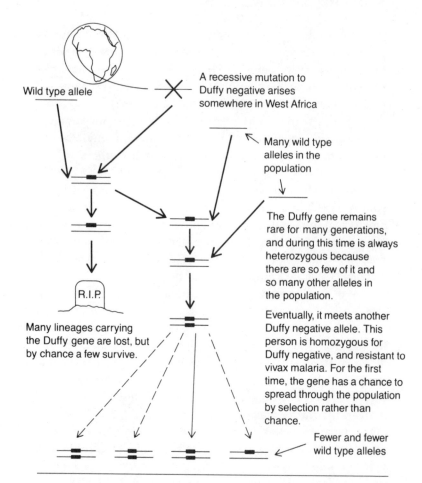

Wild type allele

A recessive mutation to Duffy negative arises somewhere in West Africa

Many wild type alleles in the population

The Duffy gene remains rare for many generations, and during this time is always heterozygous because there are so few of it and so many other alleles in the population.

R.I.P.

Many lineages carrying the Duffy gene are lost, but by chance a few survive.

Eventually, it meets another Duffy negative allele. This person is homozygous for Duffy negative, and resistant to vivax malaria. For the first time, the gene has a chance to spread through the population by selection rather than chance.

Fewer and fewer wild type alleles

Figure 1.3 The early career of a recessive Duffy-negative mutation in the West African population of many generations ago. It must have resembled the "Perils of Pauline." For many generations the gene's career was fraught with jeopardy, since it was so rare and its selective advantage could not be expressed because all the early copies were heterozygous. Indeed, it is probable that Duffy-negative arose many times before one mutant allele survived long enough to begin to spread through the population.

rica. The parasite had run out of victims. This left a vacuum, which was filled by a much worse disease—malignant, or falciparum, malaria.

Where did falciparum malaria come from? We cannot be sure, but recent studies of the DNA of these parasites show that the disease is closely related to the malarias of rodents and birds. The benign vivax malaria parasite is more closely related to the malarias that afflict our primate relatives. It appears that the ancestor of falciparum malaria may have been harbored originally in a very different species and may have adapted to live in humans quite recently. Ironically, in giving rise to a mutation that confers absolute immunity to one disease, the African population left itself open to attack by a much more dangerous disease. Genetically speaking, it went from the frying pan into the fire.

Ripples have continued to spread in the gene pool. New mutations of other genes that confer limited protection from the malignant falciparum malaria have arisen in the African gene pool and elsewhere, including the famous gene for sickle-cell anemia. None of these, however, is as effective against falciparum malaria as the Duffy-negative allele is against vivax malaria, and none of them has taken over the affected population like the Duffy-negative allele.

The neo-Darwinists realized that evolution is ultimately made up of just such ripples in the gene pool, most of them dying out as soon as they appear, a few spreading partway, and an even smaller number spreading through the whole pool and changing it permanently. As with the Duffy-negative allele, the fate of each mutation is determined by a wide range of circumstances.

The combination of mutation and many different selective agents, such as malaria, has had the effect of filling the human gene pool with innumerable alleles. Some are useful, some harmful, some both, and some have no obvious effect at the present time. To see some of the consequences of this, we can look into the future of our species.

Humans will not be around forever. If most of our species were to disappear as a result of some catastrophe, leaving only a few survivors to give rise to a new species, it might easily be that all of the current mutant alleles conferring resistance to malaria would be lost. If the malaria parasites also survived the catastrophe, the new species would be vulnerable to the disease until new resistant alleles arose.

Alternatively, all the survivors might be Duffy-negative but malaria might not survive the transition, so that the reason for the allele would have disappeared. This may not matter, but it may be that the Duffy protein is somehow important to survival in the long term, so that the new species might be at risk because it had inherited a nonfunctional gene. This has happened to us in the past. Most animals can make vitamin C, so that regardless of their diet they are in no danger of getting scurvy. But at some point in their evolution the higher primates lost a gene essential for the manufacture of this vitamin. It does not matter so long as our diet is rich in vitamin C, but it matters a great deal if it is not.

Because the future is unpredictable, a species is much more likely to survive if it is rich in genetic variation, whether inherited from its ancestors or acquired during the period of its own existence.

✓ THE PERILS OF TOO LITTLE VARIABILITY

The ripples in the gene pool are endless. Each ripple represents a new allele of one of the tens of thousands of genes. Mutation, selection, and chance have resulted in thousands of ripples that cross each other in ever-changing complex patterns. This is the reason we are all different from one another. There have never (except for the identical multiple births mentioned earlier) been two human beings with the same set of alleles of all their genes. If the human species were to persist until the stars of the galaxy die of old age, and were to increase enormously in numbers as they colonize the whole universe, the chance that two children of separate births would ever in all this time be exactly alike is still vanishingly small. Our species harbors (as do almost all others) immense amounts of genetic variability in its gene pool.

This variability is a precious resource, for it fuels the process of evolution. But if a population loses too much variability, it can flirt with extinction. This is vividly illustrated by the case of the African cheetah, the swiftest of all the land animals.

The graceful cheetah is unusually difficult to raise in captivity. Maintaining captive populations is no problem with most of the other giant cats, which can prove embarrassingly fecund. In the last few years zoos have accumulated such a surplus of lions that they can no longer be traded to other zoos and sometimes have to be quietly destroyed. This is not, alas, the case with cheetahs. Only about 10 to 15 percent of the animals captured in the wild prove capable of producing offspring, and 30 percent of these cubs die in the first few months.

Thus it was particularly tragic when in 1982 illness struck the most successful cheetah-breeding program in the world, at the Wildlife Safari Park in Oregon. The animals had contracted a viral disease called feline infectious peritonitis (FIP), which is common among domesticated cats, particularly those kept in research colonies. Even in the worst infections, no more than 10 percent of cats usually succumb to the disease, but in spite of around-the-clock nursing care eighteen of forty-two cheetahs died in the first few months. Most of the other cheetahs got sick, and all developed high levels of antibodies to the virus. Yet, as researchers at the park found, it was not a particularly virulent strain of the virus that was causing the disease, for lions that were kept near the cheetahs all made antibodies against the virus and none of them became sick.

All this and much other evidence suggested that the cheetah population was strangely fragile. Steve O'Brien and his co-workers at the National Institutes of Health, working in cooperation with zoologists in South Africa, found that by many different measures cheetahs were remarkably genetically uniform, almost clones. Perhaps their most startling experiment was to perform reciprocal skin grafts between unrelated cheetahs. When skin grafts are performed between unrelated animals, even of the same species, they are usually violently rejected in about ten days. Indeed, grafts of bits of skin from domesticated cats to cheetahs were rejected normally. But grafts between cheetahs lasted for many weeks before they were rejected; in some cases they were not rejected at all.

Other wild felines, in contrast, show normal levels of genetic variability. And until recently cheetahs themselves have been no slouches at evolving. Until about 10,000 years ago there were several species of cheetah distributed worldwide. At some point much of this precious variation was lost in the surviving cheetah lineage. This must have happened as a result of repeated reductions in population size, perhaps at the time of the waves of extinction of other large animals that accompanied the emergence of primitive man as an efficient hunter in many parts of the world. These size reductions had the effect in turn of reducing the variety of alleles in the gene pool.

This seems the most likely explanation, but for it to happen, the population would have to be squeezed through more than one very tiny-sized bottleneck. But it also means that cheetahs must have flirted dangerously often with extinction. Their precarious history makes a circus tightrope walker seem as cautious as a bank president.

Given enough time, the cheetahs might recover their genetic health. If a population of animals or plants stays large for a long time, it will accumulate the genetic variability it needs to adapt to environmental changes—this, in spite of the fact that loss of genetic variability occurs in every population, not just with cheetahs. As we saw earlier, even the most fecund of creatures fail to pass on some of their alleles in the process of reproduction. Luckily, there are usually other copies of those alleles in the gene pool. Even if you were unlucky enough to be the one to lose the only copy in the whole human gene pool of some valuable allele, you would more than make up for it by supplying new alleles to the pool.

This is because each of us, in the course of our prereproductive lives, accumulates many mutations in the cells that we pass on to the next generation. (We accumulate mutations in other cells as well. Sometimes these are important, as we will see, but they will not get passed on to the next generation.) Mutations come in a bewildering variety, and we will spend much time in chapter 4 talking about them and what causes them. They are the raw material of evolution.

Because there are so many ways a gene can be changed (remember, a gene is quite a long stretch of DNA, consisting of perhaps thousands of nucleotides), each new mutation usually results in a new allele that differs from the ones already in the pool. The overwhelming majority of these mutations are harmful to one degree or other and are quickly lost. Others have little effect, and a tiny fraction are actually beneficial. Most of these mutations, even the beneficial ones, are lost immediately because they are initially so rare. As we saw, the few beneficial ones that survive may spread all or partway through the population.

Even selectively neutral or slightly harmful alleles can spread through a population, but the likelihood that they will do so is much smaller. This likelihood is, however, increased if the population size is small, just as the likelihood that everyone in a group of people will have the same color of hair increases if the group is small. The chance that it will spread is also increased if any harmful aspects of the allele are only sometimes revealed. Cheetahs, for example, are solitary creatures, unlikely to have encountered feline infectious peritonitis in the wild. Their genetic susceptibility to this disease is conferred by specific alleles at various genetic loci. These alleles must have spread by chance through the repeatedly size-bottlenecked cheetah population, ultimately displacing any alleles that might have conferred resistance.

Cheetahs living in zoos (and that may soon, alas, mean all cheetahs) are far more likely to be exposed to this devastating disease than they are in the wild. And because they are all so genetically similar, they are all equally susceptible. It is, I repeat, dangerous for a population to lose too much genetic variability, since valuable as well as harmful alleles can be lost.

Other things can happen to the gene pool as well. Genes are reshuffled each generation by a process called genetic recombination. As a result, in every generation an allele may find itself in quite different genetic company, so that its effects on the organism may be quite different. Its effects may also change as the environment alters, as we saw with the alleles that cause susceptibility to FIP. And of course various members of a species can migrate in and out of a local population, bringing and taking an assortment of alleles with them.

The accumulation of these changes in the gene pool, introduced by mutation and sorted out by natural selection, recombination, and chance, is the essence of the evolutionary process. The result is the ten million or so species of animal and plant on the planet. But what a leap this last statement makes! My bare-bones description of evolution is the equivalent of saying that Shakespeare strung words together to write Hamlet.

The remarkable and eccentric biologist J. B. S. Haldane was one of the architects of the neo-Darwinian synthesis in the 1930s. His contributions to genetics were mainly mathematical, and along with Sewall Wright and Sir Ronald A. Fisher he founded the science of population genetics. This body of theory underpins much of our understanding of evolution, as sketched in the previous paragraphs. He was understandably fond of this intellectual construct and defended it vigorously against attacks from the Soviet Lysenkoists and from other scientists whose understanding of evolutionary theory was less profound. In particular, he subscribed to the idea that this basic bare-bones theory was indeed sufficient to describe evolution.

He called it *beanbag genetics*, contending that changes in the gene pool were essentially equivalent to taking samples of different-colored beans from a bag. The different-colored beans arose by mutation, and the process of drawing them from the bag embodied both selection and chance. As time went on, the proportions and types of beans in the bag would change, just as the proportions and types of alleles in a gene pool change. If the beanbag view of evolution is right, the collection of simple mechanisms proposed by the neo-Darwinists should be enough to explain such phenomena as the startling adaptive radiation of the Hawaiian Drosophila. Indeed, they should be enough to explain the appearance on the planet of the flies themselves, something we will deal with in chapter 8.

I must emphasize at this point that the beanbag view of evolution is indeed in its essence correct: the forces I have just listed are enough to explain the entire process. But how the forces interact and how the structures that they produce build on each other form another and equally important part of the story. Shakespeare did not write Hamlet at the age of six. He drew on the wisdom of a lifetime and the accumulated experience of his whole culture. Beanbag genetics is not the whole story. We are the product of mutation, selection, and chance, but we are also the embodiment of the accumulated wisdom of our genes.

I would like you to imagine this wisdom building in influence over time as I tell the following evolutionary tale.

EVOLUTION SOMETIMES SEEMS TO REPEAT ITSELF

Mammals, diverse as they are, have one thing in common, from which they take their name: they all provide milk to their offspring. We now know that they have had a long and remarkable history, stretching back at least 200 million years. When we examine this history we find that the same evolutionary themes often crop up again and again, to the point where it no longer seems to be a coincidence. Nowhere is this more apparent than in the saga of the mammals of South America.

The true Age of Mammals began sixty-five million years ago with the sudden extinction of the dinosaurs. The tiny mammals who were living at that time were able to fill the great vacuum that had been left by the removal of practically all the large animals on the planet.

Not long before this time South America, borne away by the spreading of the sea floor in the Atlantic, severed its last connection with Africa and became an immense island. This new island of South America eventually, about two million years ago, linked up with North America through the Central American isthmus. All indications are that South America is due to separate yet again in a few million years and take up an isolated existence once more.

At the time of South America's separation from Africa, the variety of mammals was far less than it is now. The few mammalian fossils we have from that time and earlier, unearthed in various places around the world, consist mostly of teeth and fragments of jaw.* Still, quite a bit can be inferred from careful examination of this slender evidence.

The lower jaws of mammals consist of a single bone, which

---

* These fragments are so small and difficult to find that paleontologists have had to enlist the aid of tiny helpers in the form of ants. Ants collect the teeth and bits of bone as they weather out of the rock and use them to build their tunnels and galleries. Patient sifting of the earth of anthills yields a treasure trove of these microfossils.

clearly differentiates them from their immediate ancestors, the therapsids, and also from the dinosaurs and other reptiles—all of which had jaws made up of a number of bones. Further, we can tell from their teeth that many of the early mammals were probably shrewlike insectivores, and it seems likely that most were nocturnal. One, *Purgatorius*, was similar to present-day tree shrews and might have been the ancestor of the primates—including ourselves. Others of this tiny band were apparently the ancestors of the present-day hedgehogs. And some fossil remains that actually predate the demise of the dinosaurs by millions of years resemble today's opossum, an extremely resourceful survivor of the evolutionary wars. In North America, it is the only remaining representative of a remarkable group of mammals very different from ourselves.

Marsupials like the opossum have no placenta, so that the young must be born very early in development. They complete their development inside an external pouch rather than in the womb. In the group to which we belong, the period of pregnancy can be lengthy because there is a placenta, which forms an interface between the maternal and the fetal blood circulation. Inside the placenta, tissues of mother and baby interdigitate to provide an enormous surface area over which nutrients can pass from the maternal to the fetal circulation. This permits a great deal of development to occur inside the womb. As a result of their long period of prenatal development, the newborn of many grazing animals can struggle to their feet within minutes of birth and totter along beside their mothers. Marsupials and placentals are so different that it is likely they diverged far back, perhaps as early as the beginning of the age of dinosaurs.

The mix of marsupials and placentals varies from continent to continent. Marsupials predominate in Australia in part because Australia broke off from the other continents about 135 million years ago, long before the beginning of the age of mammalian dominance when placentals began to reach their full development. There are no old placental fossils from Australia, so either placentals never arrived there or they were driven to extinction very early by the more successful marsupials.

South America presents a different history. Because it split off from Africa much later, the mammals that found themselves isolated on this new island were a more evenly divided mix of marsupials and placentals. And by this time the placentals had evolved to the point where it was possible to maintain a balance between these two groups, a balance that persisted for millions of years.

Only three types of mammals were marooned together on this vast continental island when it split off: the first was descended from the opossums and were of course all marsupials; the other two were placentals. Of these, one gave rise to the curious collection of animals that includes armadillos, sloths, and anteaters. The second, perhaps coming from Africa before the growing Atlantic split it from South America, was made up of primitive ungulates, or hoofed animals, that had begun a rapid adaptive radiation after the extinction of the dinosaurs. These animals, called condylarths, were initially browsers. But grasses appeared at about this time (surely as important an event as the evolution of the mammals), and these new plants spread quickly throughout the temperate and drier tropical zones. Many condylarths, like their counterparts elsewhere, were able to move into this new ecological niche and evolved into grazers.

The fossil record in South America is not as complete as it is in many other parts of the world, although interestingly some of the very finest fossils have been found there. A giant sloth from Argentina was the first extinct creature for which an entire skeleton could be put on display. It was sent to Europe at the end of the eighteenth century, where it attracted the attention of Georges Cuvier. Cuvier's daring and revolutionary view of the history of the planet as the product of a series of catastrophes and extinctions was based on this and on other geological and fossil discoveries.

In spite of the gaps, there are enough fossil finds to give quite a complete view of the evolution of South American mammals except in their earliest stages. Unfortunately, dating fossil finds in South America is difficult, ironically because the fossil animals are so different from those of the rest of the planet. Unless the fossils are overlaid by volcanic deposits, which can be dated by the potassium-argon radioisotope method, their dates can be quite uncertain.

It seems likely, however, that the earliest mammalian fossils found in South America date about two-thirds of the way back to the beginning of the Age of Mammals, perhaps about the middle of the Eocene. This takes us back some forty-five million years. They were entirely endemic, so that although we have no older fossil records we know that their ancestors must have been isolated for at least the previous forty or fifty million years. The immigrants who ended this long period of isolation arrived by island-hopping down from North America at the end of the Eocene, about thirty-eight million years ago. So, during this long period from about eighty-five to thirty-eight million years ago, the mammals of South America developed in what the paleontologist George Gaylord Simpson has called "splendid isolation." Indeed, even after the Eocene they remained largely isolated, since so few animals arrived from outside.

During this long period they evolved into some quite remarkable creatures. Perhaps the most interesting were the indigenous marsupials, those descendants of the opossums that had begun to radiate into various niches even before the extinction of the dinosaurs. In South America they moved into the carnivore niche, and indeed they were the only group of carnivores to inhabit South America through most of the Age of Mammals. They were driven to extinction when placental carnivores finally arrived, but until that time they had been able to prey very successfully on the placental herbivores and insectivores that had coevolved with them.

Two groups of these carnivores are of particular interest. One, the Borhyaenids (literally neighbors of the hyenas), are similar anatomically to the Tasmanian wolf, a doglike marsupial that until recently was widespread in Australia. When the placental dingo arrived in Australia between five and ten thousand years ago, the native wolf was quickly driven to extinction. (Rumors persist that a few still survive on the southern island of Tasmania.) The Borhyaenids of South America were so similar to the Tasmanian wolf that even skilled paleoanatomists have trouble telling them apart, but in fact these two animals appear to have evolved quite independently in South America and in Australia from opossum-like ancestors.

Another remarkable predator, that appears suddenly and very late in the fossil record but is also a remote descendant of the opossums, is *Thylacosmilus*, a catlike carnivore uncannily similar to the placental saber-toothed cats. A picture of its skull and that of the North American saber-tooth *Smilodon* is shown in figure 1.4.

Note the remarkably similar dentition and the projecting flange on the lower jaw of both animals. The flange must have given support to the pouches of skin that sheathed the enormously developed upper canines.

Because the history of *Thylacosmilus* is so little known, we do not know whether it was as successful as the placental sabertooths. Many different placental sabertooths evolved on other continents from the Oligocene (thirty million years ago) down to very recent times. Some representatives of these migrated into South America at about the time that *Thylacosmilus* disappeared. Indeed, their arrival probably had something to do with that extinction. But we do know that *Thylacosmilus* had evolved completely independently of the placental sabertooths. The last ancestor it had in common with its placental look-alikes lived at about the time of the

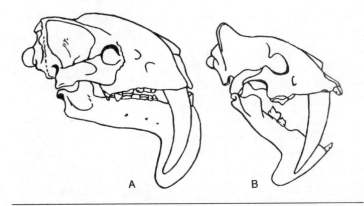

Figure 1.4 Skulls of the marsupial saber-tooth *Thylacosmilus* (*A*) and the placental saber-tooth *Smilodon* (*B*). This *Smilodon* skull is from the Pleistocene and is therefore quite recent. The *Thylacosmilus* shown here is much older, from the Oligocene. The astonishing parallel development of these two organisms is obvious. Note particularly the flange of bone on the lower jaw of both skulls, perhaps a support for pouches of skin to sheathe the enormous canines. (From figure 8 of G. G. Simpson, *Splendid isolation* [New Haven: Yale Univ. Press, 1980].) Copyright © 1980 Yale University Press.

rise of the dinosaurs, some 200 million years earlier. The remote ancestor of these two groups of mammals must have looked very different from sabertooths and indeed from anything alive today.

It can be argued that the reason the Borhyaenids and the Tasmanian wolf look so much alike is that they had a common ancestor that was already on its way to becoming wolf-like before it was split into two groups by the wandering continents. We are then forced to assume that because of the capriciousness and incompleteness of the fossil record we have not yet found these ancestors. This type of evolution is called *parallel* evolution. In parallel evolution a group of organisms splits and the subgroups continue to evolve in similar ways even though they are geographically separated.

This argument cannot be used with *Thylacosmilus*, however. Its marsupial ancestors were opossum-like at a time when the placental ancestors of the sabertooths were shrew-like. Go back to their common ancestor, and it may not even have had distinct canine teeth. Thus we have a quite certain example of *convergent* evolution, in which two or more very different groups of organisms have converged on a particular appearance, structure, or way of life.

During this long period of isolation, remarkable things were also happening to the placental mammals. As mentioned earlier, some evolved into unique creatures such as armadillos, sloths, and anteaters. But the descendants of the condylarths that had begun as browsers evolved into hoofed grazers that could run swiftly over the new open grasslands. These developed into animals that appear superficially to be very similar to some common placental families.

Two of these were the horselike and the camellike litopterns. The horselike litopterns quickly evolved the ability to run on their middle toe alone, just like horses. Some of them showed more extreme single-toed development than any other animal, to such an extent that their other toes became tiny vestiges. And, remarkably, they evolved this trait for running on firm grasslands twenty-five million years ago, twenty million years before the first one-toed real horse, *Pliohippus,* appeared in North America.

The camellike litopterns were not as much like camels as the horselike litopterns were like horses, although Charles Darwin was fooled into classifying them as camels. Their vaguely camel-like appearance was probably spoiled by a short, elephant-like proboscis. But like camels they had long necks, and their three-toed feet (camels have two toes) were splayed as an adaptation for walking on soft soil or sand.

Some of the relatives of the litopterns evolved into large, heavy-boned creatures called Pyrotheria (i.e., fiery animals, called this apparently for no reason other than an excess of poetic zeal on the part of their discoverer). They appear to have occupied the same niche as elephants today, with long chisellike tusks and elephant-like grinding teeth. Some of them became nearly as large as modern elephants, but they probably looked like nothing so much as giant, long-nosed pigs.

An observant time traveler visiting the South America of ten or twenty million years ago might have seen other animals that had become rabbitlike, rhinoceros-like and bearlike as a result of convergent evolution. These last, even though their ancestors were hoofed animals, had developed claws.

Most of these eerily familiar creatures were eventually driven to extinction by invaders from the north that occupied similar niches. The process actually took millions of years, and some of the South American animals penetrated, usually briefly but a few with great success, into the north. The few South American mammals that have managed to survive to the present were those, like the sloths and the armadillos, that were not obviously pitted against similar animals from North America.

In essence, many organisms with very different ancestries were able to move independently into parallel ecological niches in South America and in other parts of the world, and as a result evolved similar adaptations. What are we to make of this remarkable evolutionary experiment? The usual explanation for convergent evolution is that similar adaptations arise because they are the ones that provide the best solution to particular evolutionary problems, but this seems rather to insult nature for her lack of imagination. There may often be another reason, resident in the accumulated evolutionary experience of the genes themselves. It is this possibility that we are about to explore.

Convergent evolution happened both in South America and in Australia, but the convergences are more numerous and more striking in South America. For example, the grazing and browsing animals that evolved in Australia are the kangaroos and wallabies, totally unlike cattle, horses, and camels in appearance and behavior. And there did evolve a now-extinct Australian sabertooth-like marsupial, but its teeth were very different from those of *Smilodon* or *Thylacosmilus*.

One possible reason for this difference is that by the time South America became an island there were quite specific developmental patterns built into the genes of the small, primitive marsupials and placentals that lived there. These patterns were not as pronounced or perhaps as specific in the much earlier mammals living at the time of the separation of Australia. As a result of this hardwired developmental specificity, the South American mammals showed a greater tendency to evolve in the same direction as mammals else-where that carried the same patterns. Their ability to move in these directions was enhanced, and other directions were closed off to them or at least made less likely.

You will remember that all these evolutionary events were brought about by mutation, selection, recombination, and chance. Just as the neo-Darwinists said fifty years ago, these enormous changes are the summation of tiny ripples in the gene pools of all these creatures. But I think you can see that it is very hard to explain such things as the prevalence of convergent and parallel evolution or the explosiveness of adaptive radiation by the simple substitution of one allele for another. Plain vanilla beanbag genetics is not enough because genes are not structures without history. Rather, there is a higher dimension to the process: the genes have been shaped in such a way as to make certain evolutionary directions eas-ier. The mechanisms that produce this facilitation of the evolutionary process are now beginning to be understood.

2

A New Way of
Looking at Evolution?

It is more important that a proposition be interesting than that
it be true.

—ALFRED NORTH WHITEHEAD,
Adventures of Ideas

HOW EVOLVING MIGHT GET EASIER
WITH TIME

A few years ago I was walking down a busy street in Athens,
Greece, when an unusual window display caught my eye. I
found myself drawn irresistibly into an automobile showroom filled with squat, lumbering cars from another era. The
resemblance to the dinosaur room at the American Museum
of Natural History was uncanny. The showroom was empty
except for a resigned-looking salesman. The explanation was
simple: I had stumbled on a Soviet display room filled with
Volga cars. They appeared to be a cross between early Volvo
sedans and 1947 Dodges.

The salesman raised a weary eyebrow. I peered in the car
windows for a glimpse of their depressing interiors, kicked
a few tires, and retreated to the street with a sheepish grin.
And I began to think about the parallels between the evolution of the automobile and the evolution of living organisms.

Pleasingly, it now turns out that I can use these parallels both to explain what I mean by evolution getting easier with time and to illuminate the difficulties that loom in our path when we think about this way of looking at evolution.

These difficulties are considerable. A good evolutionist will be horrified by my bald statement that evolution is getting easier. But to most readers of this book, the enormity of this heading will not be immediately apparent. There is an art to making a shocking statement, and the first requirement is that it be designed to dynamite rather than merely disturb the prejudices of the reader. George Bernard Shaw said: "[T]here is no eminent writer, not even Sir Walter Scott, whom I can despise so entirely as I despise Shakespeare when I measure my mind against his." *That* was a shocking statement, designed to offend absolutely everybody. But my heading is designed to offend only evolutionists, and now I must explain why it does so.

The ideas in most books can be boiled down to a paragraph or two, without losing a great deal of meaning. Sometimes they can be boiled down to even less. Most books on child rearing, for example, can be summarized by "Don't nag." I have thought about how to be similarly succinct about this book, but have not succeeded. The reader will have to wade through quite a few paragraphs with me.

In the introduction, I unveiled the concept of evolutionary facilitation: the way genes are organized, and the way mutation-causing agents act on them, facilitates the channeling of evolution in certain directions. This is because of the long past history of both the genes and their mutation-causing agents. As certain classes of evolutionary problems have been repeatedly solved in the past, the result has been to shape in us a collection of genes (along with their accompanying mutation-causing agents) that has become very good at solving such problems.

There are two major objections to such a view of evolution. The first objection is that this idea somehow implies that organisms can anticipate new evolutionary challenges. This is obviously wrong, and explains why the bald statement of the idea has the power to shock. The second objection is more technical and concerns the kind of selection that might result in evolutionary facilitation. But both objections can be dealt with. In order to help do this, and to make matters concrete in the reader's mind, I will return often to the automobile analogy in the course of the book.

The first objection to the idea of evolutionary facilitation is obvious. Evolution does not look into the future. The whole process is automatic, driven by selection acting on individuals—in effect sorting them out and allowing only the fittest to reproduce. Evolution does not move toward some predetermined goal. There is nothing to prevent organisms from becoming simpler rather than more complex over the course of evolutionary time; indeed, nothing to prevent them from becoming so specialized that their chance of being driven to extinction by a sudden environmental change becomes very high. Nothing, including evolution, can foresee what is coming next.

There is a clear analogy with the automobile industry. Car makers exist to maximize profits. In the past, they have done so by appealing to the baser instincts of the consuming public, producing unsafe, overpowered gas guzzlers that were bought for reasons unconnected with transportation. (The anatomical implications of the front end of the Edsel have often been remarked upon.) In a climate of shortage and an era of government regulation, not only did these models become unfit, but one American car manufacturer recently nearly became extinct.

This was not through lack of planning on the part of the automakers. They spend much money and effort in trying to anticipate the new regulatory climate and the changing habits of consumers. Some companies are able to adapt more quickly than others, but this is more a function of their available resources than of any prognosticatory skills. A rich car company can build a new plant more quickly and advertise a new model more effectively. But its managers cannot see the future with any certainty, any more than you or I.

How, then, can I have the chutzpah to suggest that organisms can become better and better at adapting to conditions that they cannot foresee? The answer is simple. While the particular environmental change that looms in a species' future cannot be foreseen, there are certain general types of environmental change that occur again and again. It is these general types of change to which genes have become experts at responding.

In a similar way, car companies have become better at building automobiles per se. It takes fewer person-hours to put a car together than in the past, as robots eliminate a great deal of drudgery. Further, robots can easily be reprogrammed to perform different tasks and assemble many different models. In a way, the automotive robot is the analogue of the evolving gene arrangements that we will be talking about later in the book, arrangements that result in a more efficient organization of the genome. And the reprogramming process is the analogue of the effects of the new mutational agents that are coevolving along with the organisms they inhabit.

Automation of manufacturing has also permitted much greater quality control. I recently took delivery of a new car and found to my astonishment that there was *nothing wrong with it*. No hubcaps fell off, there were no squeaks and rattles, and all the buttons and knobs did what they were supposed to do.

Organisms in the natural world also have a history of growing expertise. We will see many examples of organisms—including ourselves—that have confronted similar environmental problems again and again during their long evolutionary histories. Such patterns are probably the norm. Luckily for life on the planet, truly unusual environmental alterations occur only rarely, and some species have by this time become so expert at evolving that they can surmount even these—though they may become dramatically altered in the process.

There is, of course, an uncomfortable aspect of this analogy with the automobile. While the short-term expertise of the car companies in improving the quality, economy, and safety of the automobile is growing, it is quite certain that the automobile as we know it is a doomed species. It has this in common with all the species of living organisms, including ourselves, that inhabit the planet at the moment. All will eventually become extinct.

The private automobile is spreading rapidly, but its day may soon be past. It has not yet reached all parts of the planet, and it is difficult to imagine it doing so. Recently I took a long trip by bus into the mountains south of the industrial city of Lanjou in western China to visit a colony of Tibetan exiles leading their centuries-old life in a remote valley. We covered about five hundred miles on very rough roads and, after leaving the outskirts of the city, did not see a single private car on the entire trip. Private cars are indeed starting to spread in China—as private enterprise has been encouraged, a few rich peasants have been allowed to buy them. But can they go on spreading indefinitely in that country of a billion people, until there are two cars in every Chinese garage? Something dramatic will have happened long before then.

That something will probably be a world climatic change brought about by our profligate burning of fossil fuel. Either the cars of the future must be powered in a totally different and nonpolluting way, or alternative modes of transportation will have to become viable. All the expertise accumulated over the last century in the manufacture of vehicles driven by internal-combustion engines will be of little avail under these new circumstances. There is no sign that the automobile companies or the car-buying public are anticipating or preparing for this inevitable change, any more than the do-

dos could anticipate the arrival of the pig and the rat on their island home of Mauritius.

Let me restate my solution to the first problem. The process of evolution is facilitated by the way that genes have come to be arranged in organisms and by the evolving nature of the factors that alter them. As a result, a species that has become an "expert" at evolving can respond quickly to most but not all environmental changes. This does not anticipate anything, but merely reflects the fact that similar environmental changes have occurred many times in the past. The shape and arrangement of our genes reflect three billion years of constant practice at evolving.

How might this have happened? It is here that we start getting into the technical difficulties, and it is here that the second objection lies.

SPECIES SELECTION AND INDIVIDUAL SELECTION

In the last chapter we talked briefly about the genomes of organisms, the collection of genes that they carry and that form the basis for the process of evolution. The automotive equivalent of the genome is the factory that makes the cars, since it holds all the information necessary to construct them. Consider the genome or factory of the Volga sedans I mentioned at the beginning of the chapter. Perhaps I am doing it an injustice, but I suspect it is not a model of efficiency. I have an image of grim sheds filled with stony-faced Stakhanovite workers bolting together parts in the same way that their fathers did. Surrounding these are immense warehouses crammed with Volga parts, the products of some five-year plan run amok, that must be exhausted before any change can be made in the cars.

The latest high-tech Japanese sports sedan, with its almost infinite variety of options, is a very different "organism." The factory that makes it—its genome—is a model of robotized efficiency. Further, Japanese automakers have cut down the stockpiling of parts to a minimum. They are ready to make changes almost immediately if a batch of parts should prove unsatisfactory or an improvement can be incorporated.

Why this difference, and is there a parallel in the natural world to these very different kinds of automobile manufacturing? Indeed there is: species selection versus individual selection. The Volga sedan is largely a product of the automotive equivalent of species selection, while the sleek Japanese machine is much more the result of individual selection.

The essence of the evolutionary process, as we saw in the last chapter, is that the fittest organisms survive and leave offspring. We will carefully skirt the quagmire that opens beneath our feet when we start trying to define fitness and say simply that on average the organisms best adapted to their environment are the ones most likely to leave offspring. They have the highest *Darwinian* fitness.

The conventional view of evolution is that there is some variation among the individuals making up the population in the amount of Darwinian fitness that they possess. As a result of this variation, evolution toward a higher average Darwinian fitness can proceed. But what about evolution of the ability to evolve? Surely this cannot happen at the level of the individual, but must happen at the level of the species, for it is species that evolve into other species. And there is a real problem with this kind of evolution.

A species is a group of organisms that have the potential to mate with each other and produce offspring. Just like the individuals that make it up, the species has a lifetime, although the lifetime may be measured in hundreds of thousands or in millions of years. Eventually, just as all organisms die, all species go extinct. Just as some organisms die without leaving offspring, some—indeed, most—species go extinct before they have given rise to new species. The dinosaurs and the trilobites are good examples of groups of species that have left no direct descendants.*

Steven Stanley, a paleontologist at Johns Hopkins University, has long been interested in questions that arise when one considers evolution at and above the species level. He is one of a group of biologists with paleontological training who tend to look at the species as the important unit of evolution. By the time biologists get to this level, they are a long way from ripples in the gene pool. Rather, they are interested in processes that are measured in millions of years, and in most cases the only way we can observe these is through the fossil record. Thus, whereas Darwin and the neo-Darwinians dealt with the way individual species members change with time, Stanley and his colleagues focus on the rules that determine how long species survive, how much (if any) they change during their "lifetimes," why some of them give rise to many other species, and why some end up without issue as collections of bones in a dusty museum basement.

We simply do not know the answers to many of these questions, but some things are becoming clear.

When one looks at the fossil record there are often long periods, sometimes millions of years in length, during which nothing much is happening. These are punctuated by much shorter periods of rapid change. _Punctuated equilibrium_—this is just what the phenomenon was named by Niles Eldredge and Stephen Jay Gould, two other members of this group of paleontologists cum biologists.

The real evolutionary action at the species level is concentrated at brief moments in geological time, when the environment, or part of it, changes dramatically. During these periods population sizes are reduced, giving the advantage to species that are full of genetic variability. It may be, as we

---

* That is, unless you count the birds, which are considered by many paleontologists to be direct descendants of one group of small dinosaurs.

will see later, that the environmental stresses themselves may trigger a higher rate of particular types of mutations. The mutational ripples spread more rapidly through these smaller gene pools. Mutations that would never have survived during times of environmental stasis, because they would be too extreme, now survive and spread, bringing about changes far larger than would take place during the times of equilibrium.

Species that cannot do this go extinct. Species that can survive.

SEX AND SPECIES SELECTION

Stanley goes a step further, and asks: What are the properties that allow a species to survive? They may, ironically, be things that do not benefit the species particularly during its long lifetime but that may enable some of its genes to survive the Götterdämmerung at the end of its career. Sex, for example.

Surely, you say, this is wrong. There must be some immediate benefit to sex. Our whole civilization seems to depend on it. We spend 50 percent of our time thinking about it. Sex redistributes the variability in the gene pool each generation, producing new combinations of genes. There has to be some advantage to all this.

Alas, no. Stanley points to a conundrum that has bedeviled population geneticists for fifty years. If an organism is beautifully adapted to its environment, then its most sensible strategy would be to make exact genetic copies of itself. If it is foolish enough to indulge in sex instead, then its offspring will have new combinations of genes. Almost all of these offspring will be worse adapted than their well-adapted parent—both because of the inevitable recombination and because the parent probably had to choose a mate with a less well-adapted genotype.

Any sexually reproducing species blunders along, made up of a ragtag collection of organisms, some well adapted and some less so, producing in each generation a spectrum of offspring with an equally large range of adaptive abilities. It survives because, ill-suited though many of its members are, some of them are the best around to fill that particular ecological niche.

But suppose a mutation arises that allows a female of the species to produce daughters just like herself without the assistance of the opposite sex. And suppose that it arises in a female with a particularly well-adapted genotype. She now has two enormous, overwhelming advantages.

First, she can produce exact and well-adapted copies of herself. No downward mobility in the offspring of this female.

Second, she and her offspring are more efficient than their sexually reproducing sisters. There is no need for them to waste half their reproductive effort producing useless males, who tend to clutter up the place, getting underfoot and drinking beer, and who do not themselves directly produce offspring. To see this, one can set up a computer simulation of a population in which such a mutant appears. If the mutant survives the first generation or two, it will take over the population with blinding speed.

This is one type of a class of reproductive mechanisms called parthenogenesis. Its advantage is overwhelming but, as R. A. Fisher pointed out over fifty years ago, it is only temporary. Asexual organisms cannot purge themselves of harmful mutations or take advantage of favorable ones as readily as can sexual organisms. And if the environment is changing in some regular way, sexual organisms can gain the advantage—though even then it is very hard for sexual organisms to overcome the twofold reproductive advantage of asexual ones. But in general, despite these caveats and if the environment is fairly constant, parthenogenesis should win in the short term.

So powerful is the short-term advantage of parthenogenesis that the only reason sexual reproduction has not disappeared may be that mutations to parthenogenesis are rare. There are no parthenogenetic mammals, for example. If it were not for this lucky circumstance, we would all be the products of virgin births.

It is only when the environment changes drastically, as it inevitably does, that sexual reproduction becomes really important again. Sexual species can quickly produce the new types needed to meet drastically changed circumstances. The result is that parthenogenetic species are much more likely to go extinct than are species that have retained sexual reproduction with its capacity to generate new genotypes.

Stanley coined the term *species selection* for this process, since it is whole species that survive or go extinct, depending on whether the species rather than the individual exhibits a particular character. While population geneticists are still arguing about it, it seems possible that species selection coupled with a low mutation rate to parthenogenesis has been enough to keep sexual reproduction around.

THE HARE AND THE TORTOISE

Other sorts of characteristics should in theory be subject to species selection, including the evolution of the ability to evolve. But there is a real problem with this kind of selection, namely that it is unbearably, agonizingly *slow*.

This is in contrast to evolution at the individual level, which can often proceed very quickly because the lifespan of individuals is short compared with that of a species. The crisis of survival for individuals strikes every generation, when they are faced with the question of whether they can last long enough and garner enough resources to be able to produce offspring. A species has a lifespan that may be measured in millions of generations, so that its crisis for survival may come only after the lapse of hundreds of thousands or even millions of years. Species selection is likely to move at a snail's pace compared with individual selection. Thus, mechanisms that facilitate evolution are likely to evolve with exquisite slowness, if at all.

Consider the Volga and the new Japanese car again. Continuous feedback from buyers and dealers drives the evolution of the Japanese car. If the car represents less than the state of the art and begins to have reliability problems, consumers will vote with their feet and buy their cars from other automakers down the street. There is little such feedback filtering back to the Volga factory. So few cars are made relative to the demand that very little buyer-driven selection can take place. The evolution of the Volga can only happen (if it does) when the warehouses run out of parts or when a directive to change the cars comes from some higher authority. The Volga will evolve very slowly as a result.

The same thing will happen to any trait in the biological world that is subject only to species selection. It is fairly straightforward to see how species selection can act to *preserve* sexual reproduction. But it is impossible to imagine how species selection might have acted to produce sexual reproduction in the first place. The evolution of something as complicated as sex demands the faster pace of individual selection. If each small advance had to wait millions of generations before it was selected, then even the unimaginably vast periods of time available during the early history of life would not be enough. This is why population geneticists continue to search, so far vainly, for convincing mechanisms that would give a short-term, individual-based, advantage to sex.

Similar arguments apply to a consideration of the evolution of the ability to evolve. It is easy to see how such evolution might happen through species selection, and much harder to see how it might happen through individual selection. It would seem that the ability to evolve should become important only when a species is threatened with extinction. Those species that can most readily evolve into a new species or group of species will survive. The others will disappear. Furthermore, during the time when conditions are fairly constant and the organisms are well adapted to them, there is no advantage to being able to evolve readily. There may even be some disadvantage: a high mutation rate, for example, would not benefit such well-adapted organisms. But this disadvantage is likely to be rather small, certainly not as enormous as the disadvantage of sex when it has to compete with parthenogenesis. So, perhaps the ability to evolve can survive long periods of equilibrium, to be expressed only during those times of environmental stress when old species go extinct and new ones are born. Those species that survive should have evolved the ability to evolve—but ah, how slowly!

Is this really a problem? After all, life has had over three billion years to get to its present enormous diversity. A good deal of selection for evolutionary facilitation would presumably have had a chance to take place even if species selection were the only mechanism operating. But I said in chapter 1 that I wanted to make the case that the evolution of all the complexity we see around us is highly *probable*. If the evolution of the ability to evolve can occur rapidly, it adds a higher-order power to the evolutionary process and makes it much more likely that complex life forms can evolve rapidly. Speed is indeed of the essence. And, just as it is impossible to imagine sex evolving through species selection, it is impossible to imagine species selection producing some of the incredibly complex genetic systems that we will examine later in this book. If our ancestors were forced to depend on species selection to produce our complex and expert genomes, it may be that three billion years was not enough.

The rest of the book will address two questions. First, has the ability to evolve indeed evolved? And second, is there some way that the power of selection on individuals can be brought to bear on this ability to evolve? If it can, then this second-order evolutionary process could occur at a rapid pace. I think it can and does happen. To see how it does requires a look in some detail at the structure of populations of organisms and at the ways in which they interact with the factors that cause mutations in them.

This, then, is the core of the present book. Not as pithy as the core of a diet book (''Don't eat so much'') or the core of a self-help book (''Do as you would be done by''), but not too bad. In the rest of the book we will explore these ideas in detail. We will see many fascinating examples of the ways in which scientists are currently studying the process of evolution, in particular, how it is now possible to probe ever deeper into the structure of the genes themselves to discover the many forces that have shaped them.

Before proceeding to the workings of evolution, I would like first to deal with some common errors in the way people think about evolution. Allowed to lie undisturbed, these preconceptions may lead to false ideas about what evolution is, and what it is not.

3

How Not to Think
about Evolution

What can we know? or what can we discern,
When error chokes the window of the mind?
    —SIR JOHN DAVIES, *The Vanity of Human Learning*

In 1981 the governor of Arkansas signed into law a bill entitled the Balanced Treatment of Creation-Science and Evolution-Science Act. The act stated that there was in fact a scientific discipline, called *creation-science*, that studied the fundamentalist Christian view of the creation of the universe from nothing a few thousand years ago. And it mandated by law that this field should be taught in the schools along with more conventional scientific views.

The bill seemed on the face of it to be a model of evenhandedness. Indeed, repeated polls of the American public have shown that a majority will say yes when the question is asked: "Do you think in the interests of fairness that the biblical story of Creation should be given equal time in the public school curriculum along with the theory of evolution?" The idea of fairness is enshrined in our Constitution and the Bill of Rights, and most people in this country profess Christian beliefs. Surely the biblical story should be given equal time along with what is after all only a theory dreamed up by a bunch of scientists.

A number of previous attempts to prevent the teaching of evolution, or at least to give the creationist viewpoint equal time, were all eventually found unconstitutional. All of them had foundered on the rock placed in their path by Thomas Jefferson: constitutional separation of church and state.

But the Arkansas act departed from all previous efforts in its simple assertion that there was no conflict, that in fact the argument was not between religion and science, but rather between two competing branches of science. (There is, of course, no doubt that what was being proposed by the fundamentalists was the teaching of a religious viewpoint in the schools.) This approach was ingenious, the result of years of careful groundwork. To bolster the act's claim that the creationist viewpoint is a branch of science, the San Diego–based Creation Research Society could point with pride to a body of practitioners of creation research. They wrote for creationist journals, published books espousing the creationist viewpoint, and held scientific meetings—just like the practitioners of any other field of science. This could all be done because the Creation Research Society is a group made up of Christian fundamentalist scientists and others who hold both strong fundamentalist beliefs and academic credentials.

These books and articles, festooned with many learned references, attacked such techniques as dating rocks and organic materials using the rate of decay of the radioactive elements present in them. In order to "prove" that the earth was only a few thousand years old, the authors would set out to show that the rate of decay of unstable isotopes was far higher in the past than it is at present, leading to dramatic errors in dating. Their books and papers were carefully designed to look like any other scientific publications, strewn as they were with complex equations.

A textbook intended for use in high schools was published. Entitled *Life: A Search for Order in Diversity*, it was actually adopted as an alternative textbook by the state of Georgia. Superficially it looked like any other biology text, with fairly accurate but quaintly old-fashioned discussions of such subjects as cell biology, anatomy, and physiology. However, the order in life that it purported to discover was not that provided by the framework of evolution, but rather that provided by the biblical account of Genesis.

The American Civil Liberties Union, unimpressed by these creationist arguments, immediately brought suit on the grounds that the Arkansas act imposed prior censorship and contravened the separation of church and state. Many religious and scientific groups filed amicus curiae briefs supporting the ACLU's suit. The case was tried over a two-week period in Arkansas superior court by Judge William Overton. Armed with their apparently scientific paraphernalia and with people whom they fondly imagined to be expert witnesses from the scientific community, the members of the Creation Research Society approached the trial with some confidence. But their "scientific" evidence was quickly destroyed by witnesses from the other side, who were able to show the gaping holes in their logic and the total lack of evidence for any of their assertions, fancy equations notwithstanding.

Indeed, the creationists' own expert witnesses turned out not to be quite what they expected. One of them was Chandra Wickramasinghe, a colleague of the heterodox cosmologist and physicist Fred Hoyle. Wickramasinghe had his own theory of the origin of life, which he cheerfully expounded to the court. According to his and Hoyle's view, life formed in galactic clouds and arrived on the earth via comets—not exactly the theory the creationists had in mind. After these and other debacles, the scientific pretensions of the Creation Research Society were effectively demolished.

The case was carefully and patiently heard by Judge Overton, who let each side have its say at great length, and who made every effort to understand the complexities of the scientific issues involved. His long and well-researched opinion was a landmark in the war between the creationists and the scientific community, which had started with the creationist victory in the Scopes trial of 1925.* Overton rejected the view that there was any scientific validity to the creationist viewpoint and affirmed that the Arkansas law had the effect of institutionalizing a particular religious viewpoint.

Overton's decision, sweeping and carefully reasoned as it was, applied only to the state of Arkansas. Recently, a similar law called the Balanced Treatment Act passed by the state of Louisiana was similarly challenged and carried to the U.S. Supreme Court, the first time that such a case has reached that level. In June 1987 the Court handed down its decision, which to the relief of biologists everywhere was seven to two against the equal-time law.

In spite of this setback to the creationist cause, the battle between creationists and evolutionists will undoubtedly continue. The courts have steadfastly blocked their attempts to force their religious beliefs on others, but that will not stop them from trying. Rejection of evolution has become a litmus test for fundamentalist belief.

It is, I think, ironic that the very system the creationists are trying to circumvent is the one that has nourished their various churches and kept them large and healthy. In America, where church and state are largely separated, a far higher proportion of the population attends church than in Western European countries still saddled with a state religion. This is because Americans are free to choose their church from the immense smorgasbord available; or they may start one of

---

* It was a hollow victory, of course. Scopes was found guilty of teaching evolution, which all agreed was against the law in Tennessee. He was, however, given only a token hundred-dollar fine. The brilliance of Clarence Darrow's defense of Scopes has totally eclipsed the unfortunate fact that Darrow was unable to get beyond the narrow question of lawbreaking and deal with the law itself. The decision against Scopes was reversed later by the Tennessee State Supreme Court, which found that he had been fined "excessively." At the same time, however, the court also found that the law was constitutional. It remained on the books—though unenforced—until 1967. For some years before that time, whenever the distinguished evolutionist Theodosius Dobzhansky lectured in Tennessee, he would announce with a twinkle that he was about to break the law. Ready though he was for the handcuffs, they were never forthcoming.

their own if they do not like any of the choices. The religious groups that filed amicus curiae briefs on the ACLU side in the Arkansas case are well aware of this. They have no desire to see a country in which people are forced to pay lip service to things they do not believe in.

The fundamentalists, for their part, are unhappy about children being taught something that their parents do not believe in. They do have an expensive and not always satisfactory recourse, which is to put their children into fundamentalist schools. But so long as these problems remain, the battle will continue.

As I noted a moment ago, the vote in the Supreme Court was seven to two, but the dissenters did not embrace the creationist viewpoint. Chief Justice Rehnquist and Justice Scalia, the two dissenters, argued that the factual basis for rejecting the law had not been aired properly as it fought its way up to the high court and that therefore it should be returned to the lower courts. The Arkansas case, in which the factual bases of both sides had been laid out in painful and sometimes embarrassing detail, was a completely separate matter in the eyes of Rehnquist and Scalia and had no bearing.

Regrettable as it is to have to admit it, Rehnquist and Scalia had a point. If the seven justices who ruled against the constitutionality of the law did so without carefully considering the arguments of both sides, then the judgment was indeed an ill-informed one. But this raises the question of how jurists, trained in the law and not in science, can decide a matter that hinges on both fact and scientifically informed opinion.

In 1897 the lawmakers of the state of Indiana nearly passed a law stating that pi had a different value from the one worked out by mathematicians. The bill had been pushed by the constituent of an assemblyman who thought he had invented a way to square the circle. Luckily, a mathematician who happened to be attending the legislative session for another purpose caught them at it, and persuaded the senate to shelve the bill (it had already passed the assembly unanimously). It was easy to show in this case that the lawmakers were unequivocally wrong.* Unfortunately, in many of the cases coming before the courts in this highly technological era, judges are being asked to make a decision, not only on legal grounds, but also on grounds of competing scientific viewpoints encompassing differing interpretations of a body of evidence. This requires that expert witnesses be called, some of whom will attempt to deceive the court.

If Rehnquist and Scalia had had their way, the equations scattered so liberally and with such superficial conviction through the pseudoscientific papers of the Creation Research Society might have had their oral counterparts in the august chambers of the Supreme Court. The justices would then have been confronted with a task for which all their judicial wisdom and training had not prepared them—to decide which of two conflicting views has the greater scientific validity.

But they could not have done this the way scientists do. Confronted by two different explanations for a particular phenomenon, scientists will search for experimental tests to distinguish between them. Indeed, the very process of formulating the different explanations often suggests which tests must be done. Thus, for example, physicists have accumulated a huge body of research that supports the theoretical expectation that the rate of decay of radioactive atoms has been constant since shortly after the big bang. Many tests of this prediction have been made, and it always holds up. But how can a lay person distinguish between this body of observation and theory and the assertion of the Creation Research Society, surrounded by pseudoscientific mumbo jumbo, that the rate of decay is not a constant?

This brings us to the heart of this chapter. How does a lay

* It is a canard, though repeated in many books, that a state legislature of the last century passed a law stating that pi was equal to three, as the Bible says (1 Kings 7:23).

person perceive the process of evolution? The lay perception ____
is obviously different from that of a trained scientist. A lay
person has not been exposed to the same body of knowledge
and certainly has not performed experiments bearing on the
problem. His or her interest in it is at about the same level
as that of any educated person in space travel or nuclear
fission or many other technical subjects.

There are, as a result, many misconceptions in the popular
view of evolution. I would like to spend this chapter on the
most common of these misperceptions—to clear the air, so to
speak, before we get to more technical matters.

THE FIRST FALLACY: ORGANISMS ARE SOMEHOW DIRECTING THEIR OWN EVOLUTION

Jean-Baptiste Antoine de Monet (1744–1829) was a naturalist
whose career spanned both the ancien régime and post-
revolutionary France. Born into a large family of the impov-
erished lesser nobility, he joined the army and fought with
distinction in the Seven Years' War. His travels from one post
to another introduced him to the natural world, and he
quickly became an excellent botanist. His early work, a massive study of the plants of France, introduced the idea of a
dichotomous key. This was a scheme for identifying animals
and plants that allowed the field taxonomist to home in on
the name of a species through a series of two-way decisions
that subdivided organisms into finer and finer groups. This
clever invention brought him to the attention of the powerful
naturalist the Comte de Buffon, who obtained an appoint-
ment for him as Royal Botanist at the Jardin du Roi. This
institution survived the Revolution (despite its reactionary
sounding name), partly through Monet's efforts. It was re-

named the Jardin des Plantes. Monet was also a prime mover in the Jardin's later transformation into the Musée d'Histoire Naturelle, a change that accompanied his own switch from botany to zoology.

But it is not as Monet that history knows him. Because of a somewhat dubious claim of his family to the seigneury of Lamarck, he styled himself the Chevalier de Lamarck. He had this order displayed prominently on all the title pages of his books. As a result, it was assumed by many subsequent commentators on his work (including Darwin) that his name was really Lamarck. And, bowing to convention, we will call him Lamarck too.

Lamarck was in many ways in advance of his time. His reputation as a freethinker aided his passage through the upheavals of the Revolution. He was, for example, one of the first to abandon explicitly the idea of mind-body dualism. He went further and claimed that human intelligence was different only in quantity from the intelligence of lower organisms.

Lamarck, like the other natural philosophers of the period, was fascinated by the apparent family relationships among animals and plants. He was led to this in the course of making many important contributions to the science of taxonomy (the naming and placing of organisms onto the branches of what we now know is a vast family tree). All his research led him to the conclusion that species had changed over time and that all organisms were related.

By the end of the eighteenth century, the discovery of increasing numbers of fossils, the finding of subspecies and varieties, and the growing evidence for the relatedness of animals through comparative anatomy were all beginning to suggest that species were not as immutable as theological orthodoxy required. In 1809, in response to this growing body of evidence, Lamarck published a book entitled *Philosophie zoologique*, which contained what was really the first explicit theory of evolution. Others had hinted at the idea, but in many cases had run into a buzz saw of opposition from the church.

He was, however, still hopelessly encumbered by the intellectual baggage of his time. His theory, as a result, was a rehash of a number of pre-Darwinian ideas about the natural world. He began by proposing that evolution was a kind of continuous Jacob's ladder, with simple organisms entering at the bottom rung and working their way up. In common with most scientists of the time, he assumed that simple organisms were constantly arising by a process of spontaneous generation. They then entered the ladder and began climbing, by gradual stages, toward man and other complex creatures.

So far so good. Indeed, this idea was not very different from the static ladder, or *scala naturae,* that was assumed by most of the scientists and philosophers of his day to order the complexity of the natural world. Lamarck simply turned it from a ladder into an escalator. While he abandoned the orthodox view that species were static and unchanging, in his theory organisms were simply working their way up through levels of complexity that had already been established, presumably at the time of the Creation.

But he then went a step further. His ideas became rather confused at this point, but he appeared to be suggesting that the motive force of this remarkable progression was the felt needs of the organisms themselves. They actually desired to become more complex. Since man was at the top of the escalator (''the most complex and perfect [organism]''), the goal of this evolution was man. He did not suggest how plants managed the feat of desiring perfection (interestingly, at about the same time, Charles Darwin's grandfather, Erasmus Darwin, wrote a long poem in which he attributed this desire to the leafy world as well).

In fact, Lamarck was forced to modify his evolutionary escalator to accommodate branching evolution and evolutionary dead ends. He also knew that the survival of organisms was important to evolution, but he assumed that their survival was directly linked to their desire to survive.

At the end of his life, Lamarck tried to apply his ideas about taxonomy and evolution to other fields of knowledge such as chemistry and meteorology, about which he knew little, and this rapidly lost him the respect of his peers. He died in poverty, lonely and ignored.

Further, perhaps in part because he was the first to stick his neck out in the matter of evolution, Lamarck's evolutionary theory was widely criticized even at the time he published it. When he thought about evolution, Lamarck seemed perversely to fall into every logical trap he could find, like a kind of eighteenth-century Keystone Kop. As a result, his name is now synonymous with every incorrect idea about evolution—including some for which he cannot even be blamed.

The most important of these latter ideas is that of the inheritance of acquired characters, which is often mislabeled as Lamarckism. While his evolutionary theory did assume that characters acquired during an organism's lifetime were passed on, he was not alone in this. Darwin, as one notable example, was forced to accept the inheritance of acquired characters in order to make his theory work, since he did not understand about genes. Where Lamarck differed, and in the process made himself a scientific laughingstock, was in his apparent assumption that organisms acquired characters by desiring them. In his most famous example, giraffes desired the tender leaves at the top of trees, and as a result their offspring had longer necks. This and other risible ideas have had the effect of totally eclipsing his excellent taxonomic work and his role in aiding the survival of French science after the Revolution.*

Indeed, his evolutionary theory had something for every-

---

* In fact, Lamarck's vague ideas were somewhat misinterpreted by later commentators. He appeared to mean that higher organisms such as man desired these changes, but lower ones such as the giraffe were driven to them by instinct. The point is a fine one, but Lamarck was not the complete idiot that he has sometimes been painted.

one to snipe at. Baron Cuvier, the early leader of the opposition, was most unhappy with the abandonment of belief in the immutability of species. Later on, Charles Darwin would criticize Lamarck both for his teleological insistence on an evolutionary goal and for his reliance on the felt needs of the organism to drive evolution. But all these fine distinctions faded after a while, and in the folklore of science Lamarck is now inseparably—and falsely—associated with the idea of the inheritance of acquired characters.

This was the idea that was put to rest by August Weismann and the early geneticists, as you will recall from chapter 1. But most of Lamarck's real ideas had been given a quiet burial long before.

Despite its manifest failings, however, Lamarckism succeeded in achieving an important goal of any scientific evolutionary theory: to remove God from the immediate day-to-day affairs of living organisms. The difficulty was that Lamarck's substitute for God was a decidedly unsatisfactory one. He had merely exchanged one supernatural force for another.

He was not alone in this. Some years ago, in an attempt to understand the eighteenth-century mind, the scientific historian A. O. Lovejoy traced in detail the religion-based idea of the Great Chain of Being, which pervaded the intellectual and scientific life of Lamarck's time. Orthodoxy stated that organisms had been fixed in this chain by the Creator. But if in fact they were free to move, surely it was only logical for them to move upward. And what more reasonable motivating force for this striving for perfection than their own desires? These ideas were part of a kind of invisible substructure of eighteenth-century thought. But when Lamarck actually put them down on paper for all to see, it quickly became obvious how ridiculous they were.

This is often, in fact, how science works. Erroneous assumptions, which we may not even realize we hold, can distort our view of the world. It is only when they are verbalized or set down on paper that it is possible to modify or abandon them. When the concept of the ether was proposed to save Newtonian physics, it led to uncomfortable contradictions. Experiments could be designed to test its existence. When that existence was disproved, the world was ready for Einstein. And Darwin might not have thought so clearly about evolution without Lamarck's straw man to demolish.

Lamarck was by no means the only one to fall into the trap of the erroneous idea that organisms direct their own evolution; the idea has persisted down to the present. The ways in which the error is expressed have become more and more subtle, however. Let me deal with just one example to show how this idea has itself evolved with time and become less glaring and therefore more insidious.

In this century a perverted brand of pseudo-Lamarckism—Lysenkoism—was responsible for the destruction of a whole science and a whole generation of scientists in Russia. This story has been told in grim detail elsewhere. But much earlier a less virulent but still damaging kind of Lamarckism had affected and diminished the reputation of a single brilliant scientist. This was Alfred Russel Wallace (1823–1913), who independently of Darwin conceived the idea of natural selection as the motive force of evolution. His brief paper on the subject catapulted Wallace to fame, for it had the effect of forcing Darwin to publish the *Origin of Species*.

Wallace was without doubt one of the most brilliant naturalists of a century that saw Darwin, Huxley, Bates, and Humboldt. Unlike most of his contemporaries in the world of science and letters, Wallace had to work for a living. His father, a Micawber-like figure, left him with very little money and little schooling. After Wallace's two great trips as a naturalist, first to the Amazon with H. W. Bates and later to the East Indies, he was forced to sell much of his scientific collection to support himself and his work. This must have diminished him in the eyes of his contemporaries, for independently wealthy and effortlessly expert amateurs, as opposed to workaday professionals, commanded even more respect in English society than they do now.

While suffering from a bout of fever in the East Indies, a sudden idea came to Wallace. The idea was one that had taken Charles Darwin years of concentrated thought to achieve, and that he was still endlessly polishing preparatory to publication. It was simplicity itself. First, some members of a species were better adapted to survive than others. Second, conditions for survival were constantly changing. Therefore, as Wallace expressed it, species should tend to depart indefinitely from their original type.

Wallace, in all innocence, wrote to Darwin and others about his blinding insight. Darwin was stunned by Wallace's apparent preemption of his ideas and actually suggested (how seriously we do not know) that Wallace should be allowed priority. Wallace was equally stunned to learn how much thinking Darwin had done on the subject and offered abjectly to withdraw his paper. Darwin's friends, particularly Joseph Hooker and Charles Lyell, worked out a compromise that had the effect of leaving Darwin's priority intact. They arranged for Darwin and Wallace to present papers at the same session of the Linnaean Society, sketches that briefly set out the idea of natural selection. The papers were essentially ignored. During the next year, Darwin rushed his book to completion, and it was the book that had the impact.

It has been suggested that because Wallace was outside the scientific Establishment, he was flattened by the Darwinian steamroller and his accomplishments were unkindly belittled. But it seems certain that Wallace, once he understood the situation, honestly felt that Darwin deserved priority. This was because of the enormous amount of evidence that Darwin had accumulated in support of his ideas, and the deep and original thinking Darwin had done about the consequences of natural selection. As will be seen, Wallace was in fact treated with consideration by the Establishment. When he managed later to shoot himself in the foot, it was without help or hindrance from anybody.

Long before he sent Darwin his letter, Wallace had placed him high in his intellectual pantheon. He had read the journal of Darwin's voyage on the *Beagle*; indeed, this was one of the strong influences that led Wallace to his own career as a naturalist. He titled his own book on evolution *Darwinism*, which is surely not the act of someone who felt that his ideas had been preempted.

Wallace's books on his expeditions sold well, his brilliance as a collector amazed his contemporaries, and he was the first to explore many virgin territories of evolutionary thought. Unlike the cautious Darwin, he rushed to apply evolutionary principles to explain the variety of human races. He did pathbreaking work on the geographic distribution of animals and plants. He discovered many examples of mimicry among the insects and suggested their probable mechanisms. And he was one of the founders of the distinguished scientific journal *Nature*. In great demand as a lecturer, he traveled to America and talked to packed houses about evolution.

Yet, his reputation has faded. In great part this is because he did not embrace the Darwinian view of the world completely. Even in his first essay on human evolution, he suggested that the evolutionary process would inevitably lead to a happier, healthier, more cooperative mankind.* This view quickly became more elaborate. In his *Contributions to the Theory of Natural Selection* (1870), he suggested that human nature and human intelligence were in fact derived from some source other than natural selection. Human intelligence seemed to

---

* So did Darwin, but he kept his thoughts under greater control. From the sixth edition of the *Origin of Species:* ''[W]e may feel certain that the ordinary succession by generation has never once been broken, and that no cataclysm has desolated the whole world. Hence we may look with some confidence to a secure future of great length. And as natural selection works solely by and for the good of each being, all corporeal and mental endowments will tend to progress towards perfection.''

him so qualitatively different from the intelligence of the lower animals that it must have been the result of some other process—ironically, quite the reverse of Lamarck's views, though it led him to a Lamarckian view of the evolution of this trait. As time went on he embroidered on this idea in ever greater detail and in more and more mystical terms. Of course, the fact that in later life he embraced spiritualism, socialism, women's rights, and other faddish notions hardly helped his reputation either.

Darwin, in contrast, hewed steadfastly to the idea of natural selection. He assumed that virtually all of evolution could be explained by its operation, though later he came to realize that chance elements could also play an important role. We saw in chapter 1 that mutation, chance events operating on the genes, and genetic population structure all play roles that were not understood by Darwin. But the basic concept of natural selection has stood the test of time. Darwin did not need to postulate mysterious "other forces" or suggest that organisms can direct their own evolution.

Unlike Darwin, Wallace was a mass of contradictions. It is amazing in retrospect that one of the two great founders of evolutionary theory should have ended up throwing so much of it away. But the whole field of evolutionary study abounds with people who have not been able, as Darwin was able, to follow this great insight to its logical conclusion.

I have spent some time on this fallacy because I want to make it very clear that what I am proposing in this book is not simply a rehash of the idea that some factor other than the basic processes of beanbag genetics is regulating evolution. Natural selection, mutation, and chance are indeed the primary shapers of the evolutionary process. But the direction of evolution, its rate, and even its likelihood have become altered with time as a result of the whole three-billion-year history that has shaped our genomes and increased the specificity of the forces that act on them.

THE SECOND FALLACY: EVOLUTION IS MOVING TOWARD A GOAL

Lamarck suggested that organisms control their own evolution as a result of their innate desires or instincts. He added to this (though his theory did not require it) that what they desired was to become more complex. At the summit of this evolutionary progression stood man.

Was man therefore the goal of evolution? He certainly was in Lamarck's view. But there is no necessity, even in an evolutionary theory in which organisms are driven by their felt needs, that these organisms move toward some goal. Even Lamarck admitted that evolution had branched, and that some of the branches had headed toward organisms less elaborate than man. It is possible to imagine a goalless Lamarckian evolution in the sense that Darwinian evolution is goalless. Because of the intellectual climate of the time, Lamarck chose not to do so. It was obvious to him that there was always a progression from simple to complex organisms, so that evolution had a goal.

Once the fact of evolution is admitted, we must confront the question of whether it is heading toward a goal. We currently know, or think we know, all the forces involved in the evolutionary process—though some aspects of these forces are still able to surprise us, as we will see. But perhaps, beyond these forces, there lurks a mysterious underlying directionality that is driving the evolutionary process.

There could, for example, be some innate imperative built into evolution that makes organisms become more complex with time. This law, proposed again and again, is sometimes called *orthogenesis*. Such a law could be imagined to operate whether or not organisms were controlling their own evolution. Perhaps it is built into the basic structure of the universe, like the evolution of stars and galaxies.

Elaborations of the idea of orthogenesis have been put forth by many evolutionists over the years. One of the most far-reaching was proposed by Louis Dollo, a French paleontologist whose career spanned the turn of the century. Dollo actually spent most of his working life in Belgium, as curator of the vertebrate section of the Royal Museum of Natural History in Brussels. He did excellent work on the ways that dinosaurs probably behaved and the evolution of the lungfish. But so far as evolutionary theory was concerned, he was, unfortunately, firmly in the tradition of Lamarck—one of a long line of French scientists who have totally misunderstood the process of evolution.

Dollo and others had observed a number of oddities in the fossil record that seemed at the time to have no Darwinian explanation at all. One of the strangest concerned the coiled oysters, *Gryphaea*, that inhabited the oceans at the time of the dinosaurs and went extinct at about the same time. These oysters, remote relatives of the bluepoints that grace the tables of those of us able to afford them, had remarkably elaborate shells. As the shells got older, each succeeding growth ring twisted slightly relative to the others, so that on older shells the boss actually appeared to curl out and over the hinge. In the most extreme development seen in some later species, the boss actually pressed on the opposing valve, preventing the two halves of the shell from opening (see figure 3.1). It is as if our mouths gradually grew smaller as we grew older, so that we had to eat progressively smaller bits of food, penultimately be reduced to drinking liquids through a straw, and ultimately starve to death.

Dollo observed such cases, along with other examples of apparent runaway development such as the giant antlers of the Irish elk, and proposed that there is an innate directionality to evolution leading to more and more elaborate types. This leads in turn and inevitably to a grotesque senescence of the species and eventual extinction.

There were enough traces of such apparent overdevelopment in the fossil record that Dollo's law was for a long time taught as a corollary to evolution. His biographer (also French)

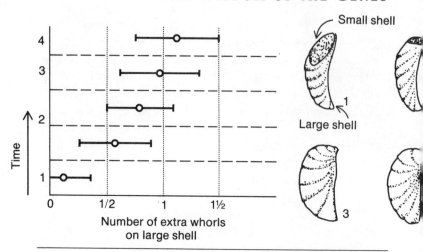

Figure 3.1 Some of the distorted shells seen in the *Gryphaea*. During the Cretaceous, in several lineages of these oysters, the larger of the two shells turned back on itself to the point where (so far as we can tell) the shells of very old oysters could no longer open. Is this an example of runaway orthogenesis, something that the French (!) evolutionist Lucien Cuénot termed "hypertely"? Probably not—these coiled shells were more likely an adaptation to give the animal some purchase in soft mud, and only very aged oysters suffered problems. (From figure 7.8 of J. R. Beerbower, *Search for the past* [Englewood Cliffs, N.J.: Prentice-Hall, 1968].)

stated that his law "enriched" Darwinism. But of course the law is nonsense.

It is obvious that any organism successful enough to leave fossils must be superbly adapted to the world of its time. By this criterion the *Gryphaea*, with their plentiful fossils extending over a hundred million years, were far more successful than most other species.

Thus, as George Gaylord Simpson has pointed out, the fact that some old shells of *Gryphaea* were closed or nearly closed off by aberrant growth at the ends of their lives tells us nothing about any underlying directional force in evolution. After all, many of us will die as a result of aberrant growths which, left untreated, would produce distortions of our bodies as extreme as those seen in elderly *Gryphaea*.

Indeed, much of the fossil record shows no sign of directionality at all. One of the first criticisms leveled against Darwin's theory was that many simple organisms had been observed to remain largely unchanged for very long periods of geological time. If natural selection is acting on all organisms, then why is it that some do not evolve, or evolve only slowly? This criticism was the result of a fundamental misunderstanding of the nature of selection, which does not have to be in some direction or other. There can also be selection for the status quo.

Consider the White Cliffs of Dover.

These magnificent cliffs are a slice through a huge chalk and limestone deposit that makes up much of the southeastern part of England. This is in turn part of a much larger deposit, some of which has been worn away or overlain by other types of rock, that stretches from the Baltic through the Low Countries to northern France.

Under the microscope, most of this chalk resolves into a coarse powder. Embedded in it, however, are myriads of skeletons and fragments of skeletons of one-celled animals called Foraminifera. On closer examination, the fragments of chalk themselves turn out to be bits of crushed Foraminiferan skeletons. The chief constituent of this whole immense deposit of material, up to a thousand feet thick in places, is the crushed remains of single-celled animals.

Foraminiferan skeletons are made of the insoluble carbonate salt of calcium, the same material that builds up scale on your kitchen taps and the deposit called calculus on your teeth. And it is the same material that gives structural integrity to your bones. The ability to build complex skeletons of calcium carbonate is very old. It evolved nearly six hundred million years ago, and when it did it led to the first explosive evolution of the animals and the first really substantial deposits of fossils.

When these crushed Foraminiferan skeletons are subjected to heat and pressure, they turn into limestone. Further heat and pressure turns them into marble. Michelangelo's David was once a collection of tiny Foraminifera living on the sea bottom.

You will recall that *foramen* means "window" in Latin. They are called window organisms because their skeletons are pierced by hundreds of tiny holes through which strands of protoplasm extend in order to snare bacteria and small plant cells. Some Foraminifera live on coral reefs, where they can grow very large. Others live on shallow ocean bottoms. And a few species, each made up of enormous numbers of individuals, spend their whole lives in the upper layers of the oceans.

Foraminifera come in a great variety of shapes, ranging from simple globes to quite complex structures reminiscent of tiny chambered nautiluses. Companies involved in oil exploration are intensely interested in fossil forams because the presence of a particular species often indicates oil-rich deposits below.

The skeleton acts as protection and may govern the rate at which the forams settle in the ocean or are moved about by currents. But it also acts as a prison. The self-contained body plan of the forams, and their mode of life, make it highly unlikely that they could ever evolve into more elaborate, multicellular organisms. That whole evolutionary direction is closed to them. As a consequence, they seem to be stuck on a kind of evolutionary treadmill.

Forams have been around for a long time. The first sparse fossils were laid down during the Ordovician, almost half a billion years ago. The first planktonic species, those that lived their whole lives floating in the ocean, appeared at about the beginning of the Age of Dinosaurs some 180 million years ago. And the greatest foram explosion occurred during the last half of the Age of Dinosaurs, the Cretaceous period. So overwhelming was the influence of these tiny creatures on the fossil record that they gave the name to that whole period of seventy million years duration, for Cretaceous means "chalky."

During that time they changed the whole face of the planet. When Foraminifera die, their soft bodies disappear like those of their relatives the amebas, but their skeletons remain. For millions of years, there was a gentle rain of foram skeletons to the bottom of the warm shallow seas that covered northern Europe. This was when the deposits were laid down that would one day be eroded to form the White Cliffs of Dover.

Then came a sudden massive extinction event, the same one that wiped out the dinosaurs. The rain of foram skeletons ceased, and eventually the shallow seas retreated from northern Europe. For a while the oceans were apparently bare of forams; then the few types that had survived gradually became numerous enough to reappear in the fossil record. Again there was an exuberant explosion of types, followed by another dieback partway through the Age of Mammals. This in turn was followed by a third great adaptive radiation that has lasted down to the present.

The striking thing about these adaptive radiations is that, while unique forms did appear each time, many forms evolved that were very similar to types that had lived tens of millions of years before. Convergent evolution is very common among the forams.

Some examples of Cretaceous and modern Foraminifera are shown in figure 3.2. You can see that, while the details of their skeletons are different, their general appearance has not changed markedly during a period of more than a hundred million years. There are many similar types, and overall the recent ones are no more elaborate than the ones from the Age of Dinosaurs.

Just as with the story of the South American mammals, there are two possible explanations for this convergent evolution. One is that certain shapes of foram skeleton are so adaptive that they tend to evolve again and again. Since we do not know why foram skeletons come in such a variety of

A

Figure 3.2 Foraminifera from (*A*) the Jurassic, the beginning of the Age of Dinosaurs, and (*B*) the Miocene and Pliocene, late in the Age of Mammals (*overleaf*). Though these fossils span at least 150 million years, and the recent ones are somewhat different in their appearance from the older ones, they

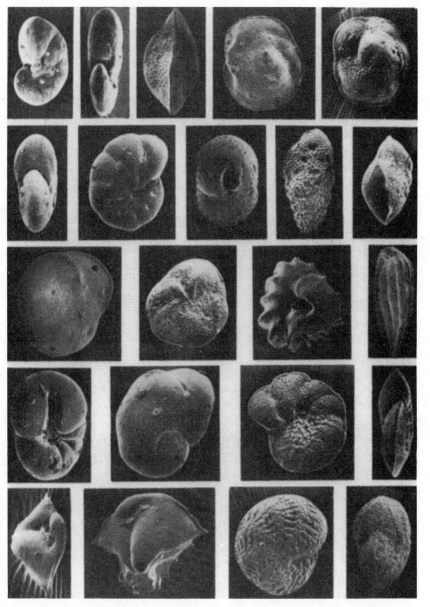

B have certainly not gained in complexity over this span of time. (Modified
   from plates 6.2.4 and 9.1 of D. G. Jenkins and J. W. Murray, eds., *Strati-
   graphic atlas of fossil Foraminifera* [New York: Halsted Press, 1981].)

different shapes in the first place, and what adaptive significance these shapes might have, we do not know whether
this is the case or not. A second possibility is that the genes
of forams are arranged in such a way as to facilitate the evolution, again and again, of certain skeletal shapes. But as
we know nothing about the arrangement of foram genes, we
do not know if there is anything to this possibility either.
Perhaps the most likely possibility is that both explanations
are true.

Regardless of the underlying mechanism, forams seem to
be on an evolutionary treadmill. Once they reached a certain
level of complexity, they could go no further. This apparent
stasis was actually used as an argument against Darwinian
evolution. It was observed that they have existed for very
long periods of time without "advancing" toward a more
complex organization. Surely they would be expected to
evolve further, but they have not done so. They would certainly have changed over time if some orthogenetic "law"
were operating.

Lamarck got around this problem, as we have seen, by
postulating that simple organisms are constantly arising to
take the place of those that have advanced up the ladder of
complexity. We now know that this is not true, and Darwin
certainly suspected it. Darwin circumvented the problem in
a more straightforward way by pointing out that if there were
no advantage to a more complex organization, it would simply not be selected for.

At one point, Darwin answered the criticisms of an expert
on Foraminifera directly:

> Dr. Carpenter seems to think that the fact of Foraminifera not
> having advanced in organization from an extremely remote ep-
> och to the present day is a strong objection to the views main-
> tained by me. But this objection is grounded on the belief—the
> prevalence of which seems due to the well-known doctrine of
> Lamarck—that there is some necessary law of advancement,
> against which view I have often protested.*

* *Athenaeum*, 25 April 1863, pp. 554–55.

Darwin had to protest a good deal against this idea because it was so pervasive and kept turning up again and again in new guises. It was very hard to stamp out the last vestiges of the Great Chain of Being.

Interestingly, these days this argument is often turned on its head. My students sometimes pose the following question: If evolution does not have a built-in direction, then why is it that more and more complex organisms *have* evolved in the course of geological time? This expresses, I think, a difference between the late-twentieth- and the late-eighteenth-century mind. Students these days are used to thinking about evolution in a nonorthogenetic way. They have been told repeatedly that there is no requirement for directionality in evolution. But then they see what appears to be directionality, and they are worried.

My answer to this new question is first to point out as Darwin did that there can be many different types of selection. As a result, organisms can be maintained at the status quo or be driven in different directions, according to circumstances. Some get more complex with time, and some, like internal parasites, can actually get simpler.

But there is more to it than that. As the complexity of many lineages of species has increased with time, so have the possible interactions among them. Living things must contend, not only with their physical environment—space, temperature, and so on—but also with other organisms with which they are competing or cooperating. This adds other layers of complexity to their surroundings. Thus, in some but not all cases, the appropriate evolutionary response to an increased environmental complexity is an increased complexity of the creatures that must deal with it.

In the course of thinking about arguments like this, it is easy to take the further step of wondering whether, since some genomes may become more complex to deal with a more complex environment, they might also become more expert at evolving.

The flip side is that some organisms such as the Foraminifera seem to have lost their ability to evolve toward greater complexity. This is a point worth pursuing in a later chapter. I should just say now that it is possible to imagine genetic mechanisms that prevent the further evolution of a particular species, or at the very least make it so difficult that it is unlikely to happen. The Foraminifera can evolve over a limited and quite rigidly prescribed range, and apparently no further. Perhaps they have become experts at going no further.

THE THIRD FALLACY: SPECIES NORMALLY EVOLVE SMOOTHLY AND DIRECTLY INTO NEW SPECIES

On a recent visit to Australia, I saw a T-shirt that showed the evolution of the Australian male. It pictured a series of silhouettes, passing from a shambling apelike creature through an upright manly fellow, then on through a sequence of ever-increasing portliness to the final stage, a large beer bottle.

We are all familiar with such evolutionary sequences, presented with grim seriousness in textbooks or museum displays, or more irreverently in cartoons. The image is so vivid that I am sure this is the way that many readers think about evolution. In fact, this is not normally how it happens. The fossil record indicates that it is only rarely that one species evolves smoothly and continuously into another. Most of the time, new species arise during periods of local or general environmental change, when the old species has been reduced greatly in numbers or fragmented into small groups.

As a result, the process of speciation usually occurs so quickly that if we are looking at the fossil record it seems as if the new species arises instantaneously. This is the common pattern that is seen even when the fossil record can be tracked in the greatest detail.

Peter G. Williamson, a doctoral candidate in geology at the University of Bristol in the late 1970s, was intrigued by these apparent discontinuities in the fossil record. The old explanation, which dates back to Darwin, is that gaps in the fossil record are due to the incompleteness of the record itself. Species are evolving all the time, so any jumps must be due to missing parts of the record, like the Nixon tapes.

But another possibility, as we have seen, is that the rate of evolution is not constant. Since the fossil record tends to compress the story anyway, periods of rapid evolution would appear as sudden jumps even if the fossil record was pretty good. Might this be true? To find out, Williamson examined one of the best fossil lineages yet found, that of freshwater molluscs from the Lake Turkana basin of East Africa.

The Lake Turkana basin is part of the Great Rift Valley, which runs down East Africa and is continuous with the Red Sea. In a few million years this whole area will be flooded and will spread farther apart to form a new ocean basin. In the meantime, it has served among other things as the cradle of mankind.

The area is now quite arid through much of the year. But whenever glaciers formed far to the north, as happened at least eight times in the last seven hundred thousand years, the climate cooled and the Rift Valley became much wetter. There have also been periods of extensive volcanic activity from a number of cone-type volcanos in the area. These intermittently covered the whole region with ash.

Wildlife was even more plentiful in the past than it is now. This is vividly apparent where streams have cut gullies through a surface unprotected by vegetation. The eroded walls of the gullies literally ooze fossil bones. The fluctuating environment and intense competition formed a kind of evolutionary crucible from which the immediate ancestors of man emerged.

Paleontologists such as Louis, Mary, and Richard Leakey and Donald Johanson have combed the eroded slopes and washes of various parts of the Rift Valley for hominid fossils. They and their co-workers have produced a detailed picture of animal and plant evolution over the last few million years in this area. The hominid fossils are the ones that hit the headlines, but they can only be understood in the context of the other animal and plant fossils found along with them. And the precision and ordering of this story would be far vaguer were it not for those volcanic eruptions that periodically spread blankets of ash over the area, like the pieces of bread in a huge Dagwood sandwich.

Fossil bones more than about forty thousand years old are too old to be dated by the carbon-14 method, for carbon-14 is a short-lived isotope and has fallen to unmeasurably low levels by that time. But the volcanic ash can be dated. This is because any rock that is heated enough to drive out gases is like a stopwatch that has been set to zero. The freshly cooled volcanic rocks and ash retain small amounts of radioactive potassium-40, which has not been driven out by the heating. As soon as the rock cools, the stopwatch begins to tick. The potassium-40 begins to decay slowly into the noble gas argon-40. This is trapped in the cooled rock and begins to accumulate. Careful measurement of the ratios of these two elements gives an accurate estimate of when the eruption took place. Thus, it is possible to date indirectly but fairly accurately the fillings between the pieces of bread in the Dagwood sandwich, fillings that consist of wind- and water-borne sediments that built up between the periods of volcanic activity. It is these fillings that contain the fossils.

Once the general pattern of dating is understood, it is even possible to date fossil-bearing sediments when the volcanic deposits have been eroded away, by looking at the animal and plant remains that are present. Each filling in the sandwich will have a different combination of species that is diagnostic. It is the tail that wags the dog in this system. It would be impossible to date the few hominid skeletons were it not for all the painful groundwork that has been done on the other and more plentiful species.

Sometimes the most informative fossils are very small. Pollen grains, which are preserved very well, tell a great deal about the vegetation of the area. Moistness or dryness can also be measured by the extent and number of "fossil" lake and stream beds, which are preserved as buried lens-shaped deposits of sand. An added bonus from Williamson's standpoint was that these beds contain numerous fossils of freshwater snails and other molluscs, often beautifully preserved. A continuous stratigraphy has been worked out for these fossils, extending back from the present to four and a half million years ago.

The great frustration faced by paleontologists is that the fossil record usually cannot be read like a book from beginning to end. It is more like a library that has been hit by a typhoon, with pages and bits of book everywhere. It can be pieced together, but inevitably there will be gaps.

There are very few places where a continuous record can be found. One of these is at the bottom of the sea, where the fossils of small one-celled organisms can sometimes accumulate undisturbed for millions of years. But as we have seen, these creatures have a rather limited repertoire of evolutionary changes. The molluscs that Williamson worked with left a record that was almost as continuous. Further, the molluscs were provided with a much greater range of characters that could be followed over time, since many different measurements could be made on each shell (see Figure 3.3).

The first striking fact that emerged from his work was that remarkably little overall evolution had occurred among these molluscs for the last 4.5 million years. Sand deposits laid down throughout this period contained a basic collection of species that showed no changes from the oldest to the young-

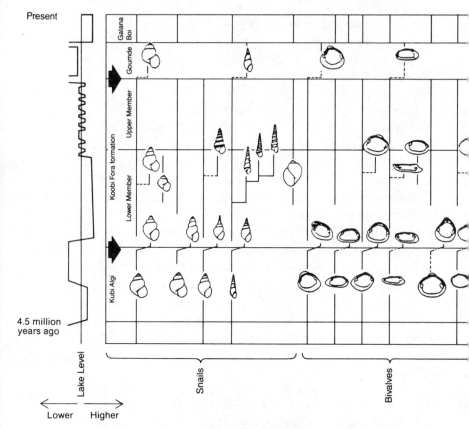

Figure 3.3 A simplified diagram of the patterns Peter Williamson found in the fossils of lake-dwelling molluscs that lived in the Lake Turkana area of East Africa over the last four and a half million years. The two heavy arrows point to violent upheavals, marked by large alterations in lake level and a temporary replacement of virtually all the species with new ones unknown both up to that time and subsequently. In each case, after a short time, the old species reestablished themselves. The older (lower) of the two events is the best documented. It is striking that the effect is seen in both snails and bivalves. Halfway between the two major events some further speciation took place, but without the accompanying disappearance of the original species. Once again, however, all the new species vanished. In spite of some difficulties with the data, this remains one of the best-documented examples of both punctuated equilibrium and the fact that most new species do not survive in the long term. (Modified from figure 4 of P. G. Williamson, Paleontological documentation of speciation in Cenozoic molluscs from Turkana basin, *Nature* 293 [1981]: 437–43.) Copyright © 1981 Macmillan Magazines Ltd.

est. Though they sometimes disappeared briefly from the fossil record, they always returned, apparently unaltered.

But the periods during which this "basic" set of species temporarily disappeared were remarkable for two reasons. First, the level of lakes in the region was changing rapidly, either falling or rising, reflecting a marked climatic change. Second, whole new sets of mollusc species appeared. They were morphologically distinct from the original basic set of species, but sufficiently close to them in appearance to be obviously related.

These new species persisted during the period of environmental disturbance, then without exception vanished when the climate stabilized again. They were replaced once more by the older set which, phoenixlike, appeared again. This meant that the persistent set of species had not vanished completely. They were always present, but were reduced drastically in numbers and perhaps displaced from many of their old haunts during the times of environmental disturbance.

Virtually all the new species that arose during the environmental disturbances were failures, in the sense that they did not join the collection of longer-lived species. Further, all these new species appeared so quickly that there is no trace in the record of transitional or intermediate types.

It might be argued that these "temporary" species were really as long-lasting as the basic set but that they lived somewhere else and were able to invade when the environment was disturbed. But there is no trace of them today. And each time the environment was disturbed, a new and morphologically distinct set of temporary species appeared. All the evidence is that environmental changes triggered a short-term evolutionary response in the form of a burst of speciation. These new species were successful in the short term but failures in the long term. This work of Williamson and that of many others suggests that the rate of evolution is often anything but constant. It can vary over orders of magnitude.

Williamson estimated that the structure of the mollusc fossil record is sufficiently detailed to distinguish events fifty thousand years apart. This is excellent for a fossil record. But of course a great deal of evolution can occur in fifty thousand years. As beanbag geneticists were quick to point out, this is plenty of time for the basic evolutionary mechanisms to bring about large changes in the gene pools of these snails. And, because of the compression of the fossil record, these changes would appear to be instantaneous.

The pattern Williamson found is unusual, not because of the evolutionary bursts, but because the same set of species kept reappearing when the original conditions were reestablished. Our own fossil record also shows discontinuities, but they are even more difficult to interpret because our record is so much more fragmentary than that of the molluscs. Our record differs from that of the molluscs in that there is no basic set of species that keeps reemerging. In our own history one species or group of species was replaced by another, and this was an irreversible event.

This is ironic, because much of our early evolution occurred in the same area and perhaps as a result of the same environmental changes that produced the repeated abortive speciations of the lake-dwelling molluscs. It appears that even though the molluscs were indeed able to evolve to meet new circumstances, the "basic" set of species quickly multiplied again when the earlier conditions were reestablished. This basic set must be superbly adapted to the most persistent East African environment. Our own evolution was under no such constraints.

Even though there is much evidence for discontinuities, we still tend to think of human evolution as the smooth gradation of one type of human ancestor into another. This is partly because the human race at the present time, despite its superficial diversity, is genetically quite homogeneous. Every person on the planet can be recognized as belonging to the same species, and we subconsciously assume that this has always been the case.

In fact, it probably has not. The recent film *Quest for Fire*, a flawed but fascinating attempt to dramatize an earlier stage of human evolution, illustrated the discontinuous nature of the human evolutionary process superbly. I thought this was the film's greatest strength. The details were probably wrong, but the idea was right and showed an excellent grasp of evolutionary theory on the part of the screenwriters.

In the film, the protagonist and his friend are apparently late representatives of *Homo erectus*, one of the last stages before the appearance of modern man. They know the use of fire but cannot produce it and must rely on finding it by chance. Their tribe has lost its only source of fire, and these two have been picked to find a new one.

They wander through a landscape peopled by a great diversity of human and near-human tribes, some more and some less advanced than they are. The tribes are constantly and viciously skirmishing with each other. Our heroes learn the secret of making fire from a girl from a more advanced tribe; she, of course, also introduces the romantic interest to the plot. There is a suggestion at the end of the film that this hybridization between tribes will produce, not only a spread of culture, but also an interesting new human type.

Despite the necessity of the film to compress and juxtapose incidents in an unlikely way, something very much like this might have happened. Indeed, the landscape of the filmmakers' imagination is not very different from the landscape reported by the early explorers of Africa in the last century. The main difference is in the extent of genetic as well as cultural diversity portrayed.

Richard Burton's accounts of his adventures in East Africa, particularly in *The Lake Regions of Central Africa* (1860), make fascinating reading. This quirky and observant traveler passed through dozens of tribes on his way to the margins of Lake Tanganyika. He meticulously recorded their divergent phys-

ical appearances and cultures and the constant warfare between them. Some tribes were much more advanced than
others, and all were very different in dress, technology, and
even appearance. Even though the Arab slave trade was already destroying the native cultures, enough remained to
demonstrate that a great deal of human diversity can be
packed into a small area.

Now consider the East Africa of two million years ago. The
diversity of hominid types and cultures must have been even
more striking. At that time at least three and possibly four
fully upright primitive hominids coexisted, perhaps even occupying the same habitats. These were *Homo habilis, Australopithecus africanus, A. boisei,* and perhaps *A. robustus.* *

Homo habilis is generally considered to be the closest of
these to our own ancestry. But it now turns out to have been
a tiny creature, perhaps only three feet tall, with long apelike
arms. Was it simply an evolutionary branch, or did it lead to
Homo erectus and then to us?

The much larger *Homo erectus,* with a bigger brain and essentially human arms, appeared in Africa over a million years
ago. It was African populations of *Homo erectus* that probably
gave rise in turn to *Homo sapiens* one or two hundred thousand years ago. We used to think that *Homo habilis* was the
ancestor of *Homo erectus* and, therefore, the grandparent of
humankind, but we are now no longer so sure.

We may never know the complete story of what happened,
but it is quite certain that there was a great deal of evolutionary action in the Rift Valley of two million years ago. This
has had to be inferred from the slimmest evidence. The fragmentary fossil finds of all these hominids could easily be
placed on a dining room table. We can tell from scratches on
their teeth what their diets probably were. For instance, *Australopithecus robustus* was small-brained with huge grinding

---

* The fossils of the large and strong *A. robustus* have been dated somewhat later, but fossils tell us nothing about when an organism first appeared,
only when it flourished.

molars, and the scratches on them suggest that its diet was primarily vegetarian. Yet it may have had some very human characteristics. The shape of some hand bones of this creature, recently found in South Africa, suggests that it was able to use tools. But of course we do not know whether *A. robustus* really did use tools at all. And we do not know in particular whether the few examples of what may have been stone tools from strata of the same age were actually used by *A. robustus*.

These scattered bones will never tell us whether these various putative ancestors of ours painted their faces, worshiped gods, sang, or danced. It seems quite certain that their behaviors were clearly different from each other, and that this may have brought them into conflict. And these conflicts may have spurred their evolution.

We know our own ancestry better than we know that of most other organisms. But in spite of all the effort expended to fill them in, huge gaps still remain in the record. Back beyond a million years or so we have no clear idea of which (if any) among all those competing creatures were our own ancestors.

Discontinuities in the fossil record seem to be very common. But when we study a process as diverse and complex as evolution, we begin after a while to expect to find exceptions. And indeed, it turns out that there are a few examples from the fossil record in which an entire species appears to pass by imperceptible stages into another. Many of these fossils come from deep ocean cores, where long periods of evolutionary time can be studied without interruption. Here the evolution of the one-celled planktonic organisms, the Foraminifera, diatoms, and Radiolaria, can be followed without interruption. But examination of these lineages also points up clearly the difficulties we have in interpreting the fossil record.

One of the best of these sequences was found by Björn Malmgren of the University of Uppsala. He examined cores taken through bottom deposits from the south Indian Ocean and followed the fate of one species of our old friend, the Foraminifera. He was able to follow all the stages of its change into what was apparently another species. This foram has a shell that resembles a chambered nautilus, a flattened coil. From the edge it looks like a discus. The cores allowed him to follow changes in the size and shape of the shell over the last ten million years.

For the first four and a half million years there were significant fluctuations in the size and thickness of the disc, but no particular trend. Then, over a half-million-year period, between about five and a half and five million years ago, it rapidly but smoothly changed from a small, slim disc to a large disc that was proportionately much thicker in the center. This is depicted in figure 3.4. Because of this change, it was given a new species name. Since that time this new species has remained fairly static in appearance. There have been significant short-term fluctuations, but it has always returned to the same size and shape.

In this case all the transitional forms are present, like the Australian evolving into the beer bottle. And many specimens of all the transitional types are found in the core, so there was no obvious reduction in population size at any point. We find in this lineage pronounced differences in the rate of evolution over time, with most of the evolutionary change packed into a (relatively!) brief half-million years. Of course, we cannot be sure that these are true species. The recent form looks like a larger, fatter version of the older form. It has been given a new species name, but since we know nothing about the genetic differences between the two species, we do not know whether the two groups really were reproductively isolated as the biological definition of good species demands.

It is interesting that all the cases of gradual evolution that we know about from the fossil record seem to involve smooth changes without the appearance of novel structures and func-

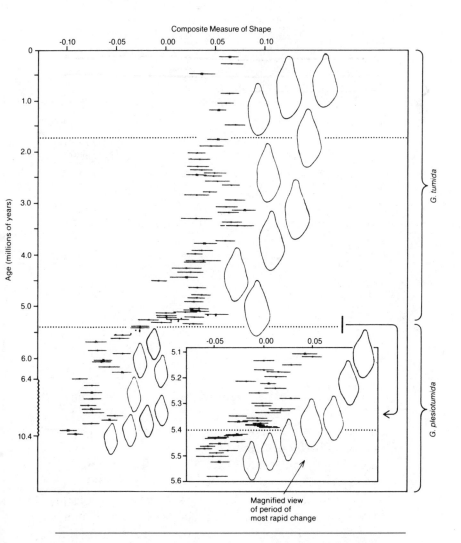

Figure 3.4 A detailed survey of the way one Foraminiferan species changed into another over a half-million-year period some five million years ago. Two things are unusual about this fossil record. First, unlike most fossil records, it is continuous. Second, plentiful fossils of all the intervening stages between the two species have been found. Are these really two different species? We will never know, but there are certainly true genetic differences between them, differences that were produced gradually. (Adapted from figure 1 of B. Malmgren et al., Species formation through punctuated gradualism in planktonic Foraminifera, *Science* 225 [1984]:317–19) Copyright © 1984 by the American Association for the Advancement of Science.

tions. It may be, as a number of researchers have suggested, that really dramatic changes can occur only during the violent alterations in the gene pool that happen most readily in small, ephemeral populations.

To describe speciation properly it is necessary to deal with both the fossil record and the genetic mechanisms of species formation. Both usually exhibit discontinuities. The discontinuities in the fossil record are physical. The discontinuities in the genetic process of species formation result from strong selection acting on a small sample of the original gene pool of a species. One cannot be considered without the other, and taken together they show that there are many possible avenues for speciation.

THE FOURTH FALLACY: THE FOSSIL RECORD GIVES A COMPLETE PICTURE OF EVOLUTION

From what we have seen already, we would expect the fossil record to be very biased towards organisms that live in large populations and do not do much detectable evolving over long periods. Indeed, the Foraminifera provide a vivid example of this. Although they are single-celled, forams, particularly the bottom-dwelling ones, are mostly quite large, easily visible to the naked eye. About twenty skeletons of an average-sized species can be packed into a cubic millimeter, about the size of a pinhead. Because they are hollow, thin-walled, and easily crushed, one could probably squeeze ten times that many into a pinhead after crushing and packing them. Just for fun, I have done some rough calculations.

There are at least 50,000 cubic kilometers of Cretaceous chalk deposits in southern England, northern France, and the adjacent English Channel and North Sea. A millimeter is one-millionth of a kilometer, which means that a cubic millimeter is one-million-million-millionth of a cubic kilometer. A more manageable way of saying the same thing is that one cubic kilometer contains 10^{18} or 1,000,000,000,000,000,000 cubic millimeters. The total number of Foraminiferan skeletons, then, is 200 times 50,000 times this number, or 10^{25}, a one followed by twenty-five zeros. And most of these skeletons fall into a relatively small number of species.

This is an enormous number. There are about five billion, or five times 10^9, human beings on the planet at the present time. Five billion Foraminiferan skeletons could be packed into a cube less than thirty centimeters on a side, which is less than a cubic foot. Foraminifera are large for single-celled organisms, but compared with human beings there are a great many of them.

As a consequence, any signs of splitting and rapid evolution of small populations of these organisms will be swamped by the enormous numbers of the more successful species. Foraminiferan deposits, while ideal for following gradual evolution, are not the place to examine speciation in all its aspects.

Other fossil records have the reverse problem. I mentioned that all the East African hominid fossils that have been discovered could fit on a dining room table. And while no self-respecting natural history museum is without its dinosaur skeleton, only about five thousand fairly complete dinosaur skeletons have been found. (Lots of fragments turn up all the time, but the finding of a well-preserved skeleton is something of an event.) This is rather a pitiful collection of remains for a group of organisms that was the dominant life form on this planet for over 120 million years.

But, like the *Gryphaea* I talked about earlier, the dinosaurs have done better at leaving fossils than we probably will. The first certain examples of true modern humans, the famous Cro-Magnon man, lived about twenty-five thousand years ago in southern France. Molecular evidence suggests that we originated considerably earlier, in Africa, but there is no inarguable fossil record of this. And the molecules give no hint of when we made the transition (if we indeed did) from *Homo erectus* to modern man.

Let us imagine for a moment that twenty-five thousand years ago some catastrophe wiped out our species. And let us imagine further that some fifty or a hundred million years hence a new intelligent species arises and develops a science of paleontology. If they begin hunting for signs of intelligent life in the past, will they find traces of *Homo sapiens*?

Highly unlikely. By that time, all the East African fossils will be part of the continental shelf of a new ocean formed through the widening and flooding of the Rift Valley. The European and Asiatic fossils of *Homo erectus*, Neanderthal man, and modern man have been found mostly in caves or in shallow burials. These would certainly have disappeared over a span of millions of years. The entire history of the development of modern man, down to the time when culture began multiplying our numbers, would have vanished without a trace.

I often wonder what remarkable similar stories have vanished, as a result of the incompleteness of the fossil record.

This incompleteness is expressed in other ways. By far the most numerous higher organisms on the planet, both in terms of numbers and in terms of species, are the insects. They have swarmed over the earth since long before the dinosaurs, and their ancestors were probably among the first animals that colonized the dry land. Yet, although they have hard skeletons, very few of them have been preserved. This is because their skeletons are not bone or limestone or glass like those of easily fossilized organisms, but rather a mixture of protein and carbohydrates. These digestible skeletons are quickly destroyed by oxidation and bacterial action.

By far our largest collection of insect fossils has been found beautifully preserved in amber. Immense deposits of amber were dug up during the last century from the North German seashore. The amber is hardened pine resin, in which the insects became mired while it was still liquid. Lumps of this resin were washed down rivers, now long vanished, that had their source in great northern European pine forests. Eventually the lumps of amber were deposited in estuaries. The estuaries themselves became fossilized as the rivers moved with time or dried up.

Much of this precious collection ended up in Berlin. Other extensive amber deposits have been found in North and South America, but none so rich. And almost all the Berlin collection was destroyed by Allied firebombing during World War II. Within only a few years, we have succeeded in finding and then destroying most of the fossil remains of the largest group of organisms that has ever lived on this planet.

It is not surprising, then, that the fossil record is incomplete and discontinuous. What is surprising is that we have been able to discover and to infer so much from this fragmentary record about organisms that lived in the past. Let me just give a few brief examples of the success of this paleontological detective work.

We now have evidence that many of the dinosaurs may have been warm-blooded, that some of them took care of their young after they hatched from the egg, and that some of them had well-articulated "hands" and opposable thumbs. Not all the brains of dinosaurs were tiny. And it is now almost certain that a giant meteorite really did crash into the earth about sixty-five million years ago, though whether this was what really caused the extinction of the dinosaurs or whether it simply helped the extinction along is still being debated.

This controversy has been fueled by the discovery that just before the meteorite impact mammals were becoming more numerous and diverse while the diversity of dinosaurs (though not their numbers) was diminishing. It seems that the mammals were already nipping at the dinosaurian heels, and that the arrival of the meteorite may have been the coup de grace.

Mammallike creatures roamed the earth in great numbers well before the Age of Dinosaurs. Some of them were undoubtedly the ancestors of the present-day mammals, and I will tell their story in more detail in chapter 10. They were very different from present-day mammals, but their similarities were great as well. A brisk debate is going on about just how different they were and how much early mammalian evolution may have taken place during that distant time.

Fossils also allow us to infer a great deal about the ancient climate of the earth. There were times when massive glaciers ground their way from the poles down almost to the equator, causing many of the organisms living at the time to go extinct. There were other periods when there were no glaciers at all, and much of the earth basked in a tropical climate. During the Carboniferous, a time of extensive rain forests, the air was so rich in oxygen that giant dragonflies with two-foot wingspans could buzz among the swampy vegetation.*

But it is exceedingly rare to find a beautifully graduated series of fossils that show exactly how one species evolves into another. This would probably still be the case even if the fossil record were much better than it is.

In spite of that failing, the battered fossil record still manages to tell us a great deal about evolution. As we have seen, even the fact that it is discontinuous tells us something about how species evolve. And, to anticipate a little, it can also tell us something about how the process of evolution has itself evolved.

I hope the reader who already knew all this stuff has borne with me through this chapter. And I also hope that it may have clarified for some readers the nature of some common

---

* Recent reports that the air was enriched in oxygen during the much later days of the dinosaurs, based on measurements taken of air bubbles trapped in old amber, appear to be erroneous.

misconceptions about how evolution works. Now it is time to see how evolution really does work.

A hundred years after Darwin, our ideas about evolution are still changing rapidly. This is nowhere so apparent as in the study of mutations, the sudden alterations of genes on which all of evolution depends. The process of mutation is not as simple as early geneticists once thought, or as the general public still thinks. As we learn more about it, it is turning into one of the most fascinating and complex processes in biology.

4

The Role of Mutation

Species and the higher macroevolutionary steps are completely new genetic systems. The genetical process which is
involved consists of a repatterning of the chromosomes which
results in a new genetic system. The theory of the gene and
of the accumulation of micromutants by selection has to be
ruled out of this picture.
— RICHARD B. GOLDSCHMIDT, *The Material Basis of Evolution*

THE RAW STUFF OF MICROEVOLUTION

It may have been somewhere in the Great Rift Valley that
our ancestors first discovered the virtues of fermentation.
Fruit juice, or perhaps a dilute solution of honey left in a
covered pot for a few days was found to bubble and foam.
When the brew was drunk, it produced a most pleasing effect
that added an exciting new dimension to tribal ceremonies
and helped to soften the considerable rigors of day-to-day
existence.

It took a million years or so to progress from this undistin-
guished fizzy stuff to the infinite subtleties of Chateau Mou-
ton Rothschild. And it took that same million years to discover why the process of fermentation actually takes place.

We now know that yeasts, simple single-celled organisms,
are everywhere in the environment. When a yeast cell falls
into a sugar solution it begins to multiply furiously. If there
is little or no oxygen around, its numerous progeny produce

alcohol and carbon dioxide as waste products. It is the alcohol that we normally think of as the product of fermentation, but the carbon dioxide has many uses too, notably in the leavening of bread flour. And the alcohol plays an important role in the tiny ecology of a fermentation vat, for it prevents the growth of competing organisms.

But if we now shake or agitate this fermenting mixture, so that oxygen is introduced, the yeast is able to convert the alcohol from a waste product into a foodstuff. As it uses the alcohol up, other organisms can begin to grow. Our wine can turn to vinegar or worse.

Both the last step in the production of alcohol and the first step in using it up are catalyzed by an enzyme, called alcohol dehydrogenase. A few years ago, our group at San Diego began studying mutants of this enzyme. We had two reasons for doing this. First, we wanted to try to improve the way the enzyme worked. Second, and more important, we had a really clever way to make mutants, and we wanted to see what would happen when we tried it out.

Like most enzymes, alcohol dehydrogenase is a huge protein, made even larger because each of the enzyme molecules is a collection of subunits. The functioning molecule consists of four identical subunits, loosely bonded together. Each of these in turn is made up of 347 amino acids. And each amino acid can be fairly elaborate. So you can see that the complexity of the molecule is enormous, as is its potential for change.

A single gene codes for this enzyme subunit. When enough subunits have been made from the instructions coded in the gene, they start to come together in quartets inside the cell. Once they have done this, they can begin to function—unless a change has occurred in the gene. If a mutant rather than a wild type subunit is made, it often may not function or may function improperly. Other mutational changes will have no measurable effect on the molecule. And some may actually improve it.

In fact, of course, the term *improve* is highly subjective. The human experience in selecting ''useful'' organisms demonstrates this vividly. Human chondrodystrophic dwarfs have a defect in the development of the long bones of the arms and legs, so that their heads and trunks are normal but their arms and legs are very much shortened. The defect can arise from a mutation in one of several genes.

Chondrodystrophic dwarf mutations very similar to those in humans have appeared spontaneously in a number of breeds of dogs. The most extreme of these has resulted (after a good deal of subsequent selection) in the dachshund. Back in the days when it was first bred, this dog was much prized for its ability to squeeze down rabbit holes and other tight places. But was this an improvement over a normal set of limbs? For the dog breeder and the hunter, certainly, but not for the dog.

When our civilization comes to an end, some breeds of dog may survive. But not the dachshund.

Similarly, the clumsy efforts in our lab were hardly likely to make any real improvement on a molecule that has been honed by billions of years of selection. But we thought that we might, in our tinkering, make an enzyme that could produce ethanol more quickly. This would be an advantage not to the yeast but rather to the brewmaster or wine maker using it.

Each of the long chains of 347 amino acids in the protein is coded by a triplet of bases in the DNA of the gene. This means that the gene is three times 347 or 1,041 bases long. This is fairly small as genes go. But there are lots of ways to change even a small gene.

Pretend for a moment that the gene consists of 1,041 colored snap-on beads locked into a chain. The beads come in four colors. These correspond to the four different bases in the DNA: adenine, thymine, guanine, and cytosine (but it is probably easier to think of them as beads colored amber, turquoise, green, and coral).

Because all the information is coded along one chain of DNA's double helix, there are no constraints on the order of the colors. Any bead can be followed by another in any of the four colors. Of course, the order of the beads does put a constraint on the order of the colored beads that run along the other DNA chain, so that the double helix can make copies of itself properly, but that does not concern us here.

A sequence of three beads codes for an amino acid. How many different sequences of three beads can there be? Any of the four colors can be in the first position, any in the second, and any in the third. So there are $4 \times 4 \times 4$, or 64 possible sequences, ranging from three ambers in a row to three corals, with all the combinations in between.

But it turns out that there are only twenty amino acids specified by the genetic code. This discrepancy was explained by the discovery that most of these amino acids are coded by more than one sequence. And 3 of the 64 sequences code for "stop," which brings the protein to an end. The result of all this is that if you were to substitute a bead for another of a different color, you would produce an effect on the protein only a little more than two-thirds of the time.

Suppose we have the sequence amber, coral, green. Putting turquoise in the place of amber will change the amino acid (from threonine to serine in this case), but putting turquoise in the place of green will have no effect. In spite of this small restriction, the number of possible proteins that can be built out of this system is enormous. And the number of possible changes that can be made in a protein once it has evolved is almost as great.

The techniques of molecular biology have allowed us to investigate some of these changes. It is now actually possible for workers in the laboratory to make any alteration that they can think of in the DNA of a gene, ranging from the substitution of one color of bead for another to the deletion or insertion of a substring of beads and even to something as dramatic as duplication or inversion of a section of the string.

Each change still requires a good deal of work, though with each passing year the techniques involved become easier and more precise. And this opens up such an astronomical number of possibilities that it is difficult to know where to begin.

Consider just the changes we could make in our alcohol dehydrogenase gene if we wanted to substitute one amino acid for another. We could, after some manipulation, remove any three-bead coding sequence and substitute another for it. Thus, we could make mutants in which any of 19 alternative amino acids could be substituted for the original one at each of the 347 places on the molecule.

How many different ways could we do this? If we consider just the molecules in which just one amino acid has been changed, the number of possible substitutions would be 19 in the first position, 19 in the second position, and so on. This gives 19 × 347, or 6,593. It might be feasible, given enough time, to investigate all these changes and find out which if any of them was "better" than the original.

But these timid evolutionary steps do not begin to exhaust the potential for change. The number of possibilities goes up dramatically if we consider all the proteins that differ from our original by two amino acids. Each of our 6,593 proteins could now have a substitution in any of the remaining 346 positions (the ones that have not yet changed). And the substitute could be any of the 19 amino acids not currently found at that position. The number of possible proteins that differ by two amino acids from our original is given by 6,593 × 346 × 19, or 43,342,382. It would not be feasible to investigate all these proteins, or even a significant subset of them. Take the argument a step further, and we get 43,342,382 × 345 × 19, or 284,109,314,010 different possible ways in which we could build a protein that differs from our original by three amino acids.

But we already know that most of these substitutions and combinations of substitutions would be extremely uninteresting. Many of them probably would not do much to the molecule, and many others would destroy its function completely. Such mutations would certainly play no role in evolution. The secret of the power of natural selection is that, while many different substitutions can arise in each generation, only a tiny fraction of them would survive and an even tinier fraction might produce an enzyme that was in some fashion improved. In each generation natural selection winnows out most of the new mutations, leaving this tiny fraction of a tiny fraction. Even the most indefatigable molecular biologists would be unlikely to stumble on any of the useful mutations by simply plodding through all the permutations.

We wanted to look at this evolutionary process, or at least some approximation of it, in the laboratory. To do this, we had to ensure that the mutations we made in the gene were the ones that might be selected for during the course of adaptation to new conditions. Otherwise, we would end up screening immense numbers of cells and the few mutants we found would probably be rather dull. We also wanted to make sure that the mutations would do something interesting to the enzyme. We did not want to pick up mutants in which the enzyme was simply wrecked.

Thus, it was necessary to come up with a way to select for mutations within the gene itself that would adapt the yeast to new conditions without destroying the enzyme's function. This was possible because we had earlier discovered a way to back the cell into a genetic corner from which there was only one escape.

Yeast cells can live in the absence of oxygen. Even if the part of the cell that uses oxygen is destroyed, the cells can go on multiplying (and of course making alcohol and CO_2). They grow more slowly than before, but they survive very nicely.

We found that when the yeast cell is disabled in this way, and no longer able to use oxygen, it is absolutely dependent on its alcohol dehydrogenase for survival. If it loses the function of this enzyme, it dies. We then supplied these disabled cells with a different alcohol, one that could be changed by the enzyme into a poisonous compound. The mutants that survived were resistant to the potentially poisonous alcohol because they had an altered enzyme. Their enzyme could still act on the dangerous alcohol, but it did not convert as much of the alcohol to the poisonous compound as the wild-type enzyme did. In this selective scheme, mutants in which the enzyme was simply destroyed did not survive.

All the mutants we obtained were the result of the substitution of new amino acids for old ones at various points in the protein. One of these substitutions was particularly interesting. We found that it had inserted an amino acid into the yeast protein that was the same as one that is ordinarily present in the corresponding place in the alcohol dehydrogenase of the horse. As a result, this substitution made the yeast enzyme behave more like the horse enzyme. This in turn had the effect of making the cell more resistant to the dangerous alcohol.

This is not to say that horses are more resistant to the dangerous alcohol than yeast. But, serendipitously, the horselike change in the yeast enzyme had the effect of making the cell carrying the change more resistant to the alcohol. From an evolutionary point of view, it was fascinating to see that we had actually retraced a path that nature had taken. The joke around the lab was that we had made the first step in the long and painful process of turning yeast into a horse.

In a way, we had. We had done in miniature what has happened untold billions of times in the course of evolution. We were studying the fine structure of evolution. Such tiny events are referred to as *microevolutionary* changes, as contrasted with the *macroevolutionary* changes that lead to clear and obvious differences between species.

Our microevolutionary change was not a random change in the gene. This particular amino acid plays a very important role in the way the enzyme works, and it is not surprising that one of our changes affects this critical part of the molecule. But in spite of its importance, it had no effect on the appearance of the yeast cell. How can we get from such tiny changes to alterations that we would call macroevolutionary? After all, it is immediately clear to the most obtuse observer that there are pronounced differences in appearance between horses and yeast. How might such large and obvious changes have accumulated in the course of evolution? They were certainly produced by mutations. What kinds of mutations?

We could have repeated our experiment over and over, and perhaps eventually have persuaded yeast to make a very horselike enzyme. But it is obvious that we cannot get from yeast to horse by simply repeating this process indefinitely. There are huge differences in the gene organization of horses and yeast. This can be dimly glimpsed from the fact that horses have over two hundred times as much DNA per cell as does the lowly yeast.

In chapter 1 we mentioned beanbag genetics, which proposes that macroevolutionary changes are simply the sum of large numbers of microevolutionary alterations. Most evolutionists agree that this is true. But we know that amino acid substitutions are not the only possible kind of microevolutionary change. If large macroevolutionary changes are to occur, there must be more to microevolution than simple amino acid substitutions.

This in turn means that many different kinds of microevolutionary changes, summing up to macroevolutionary changes, must have happened in the course of the five billion years or so of evolution that separate horses and yeast from their common ancestor.* In this chapter we will examine some of the kinds of changes that must have occurred, most of which are much more substantial than simple substitutions

---

* Why 5 billion, when the earth is only 4.6 billion years old? Remember that yeasts and horses, in spite of their obvious differences in complexity, have equally long evolutionary histories. If their last common ancestor lived 2.5 billion years ago (plus or minus half a billion or so), then they are in effect separated by twice that period of evolutionary time.

of one amino acid for another. Many such different kinds of changes can happen, and there are undoubtedly many that we do not yet know about.

MICROEVOLUTION AND MACROEVOLUTION

Some years ago I had the doubtful pleasure of joining a small band of evolutionists in a debate with members of the Creation Research Society. The debate took place before a vast audience consisting primarily of students who had been bused in from a nearby Christian university. The creationists dazzled the audience with slides such as one showing human and dinosaur footprints overlapping each other (this particular find has since been officially debunked). Judging by the thunderous applause at the end, they appeared to win the debate hands down, as they usually do in such a forum.

The impact on the audience, however, had been minimal. We know this because, at the end of the debate, the moderator asked whether anyone had changed his or her mind as a result of what had been said. A solitary hand went up far in the back. But the owner of the hand was swept away by a tide of people heading for the buses before we had a chance to find out *which way* his mind had been changed.

What I found most interesting in the debate was not the appalling collection of pseudoscientific gibberish trotted out by the creationist side or the gullibility of the audience. Rather, it was that Duane Gish, the articulate spokesman of the creationist team, handicapped his side at the very beginning: he accepted microevolution. Yes, he said, Darwin's finches on the Galapagos Islands were an excellent example of microevolution. He just happened not to believe in *macroevolution*. (The other two members of the creationist team would not concede even microevolution. They felt, I think quite properly, that if one didn't believe in evolution, one had to go all the way.)

The cynical interpretation of Gish's stand, and I suspect the correct one, is that he expected the evolutionist side to trot out lots of well-documented examples of microevolution. By conceding this point, he could shift the grounds of the debate and do what he does best: ridicule scientists' claims about the large changes brought about by macroevolution.

But it turns out that many scientists not motivated by religious considerations or a hunger for power are also worried about the gap between micro- and macroevolution. Is one simply a summation of the effects of the other? Are the large differences we see between species entirely due to small genetic changes such as our yeast-to-horse amino acid substitution? Or is there some other process at work?

I hasten to note that nobody is resurrecting the kinds of orthogenetic processes we talked about in the last chapter. Rather, some new kind of mutation is envisioned, one that can bring about a large genetic change in one step. If such mutations were to be accepted by natural selection, they might not only speed up the evolutionary process, but they would also produce the kinds of large qualitative changes that tiny mutations seem unlikely to be able to do.

This notion begins to hint at the ideas that are at the core of this book. If there are really two kinds of mutations, small ones that can be made and studied in the laboratory and large ones that happen occasionally and greatly speed up evolution, then the large mutations might be just what is needed to facilitate the evolutionary process. Do such large mutations occur, and what might they look like?

This idea of large mutations, like most ideas about evolution, has a long history. It goes back to the end of the last century.

As a boy in Holland in the 1860s, Hugo de Vries grew up in a privileged family. His father was minister of justice in the Dutch government, and both sides of the family tree were festooned with distinguished relatives. But he ignored the beckoning fields of law and statesmanship to concentrate on plants. By the time he finished his years at the gymnasium, young Hugo had made a complete collection of all the Dutch flowering plants.

In his subsequent career he went on to do distinguished work in plant physiology, but his early reading of Darwin had turned his thoughts to the mechanism of inheritance and how it bore on evolution. Like Mendel, but quite independently of him, he began to make crosses between lines of plants that differed in single characteristics. Unlike Mendel, he did not make crosses between lines that differed in more than one character, or carry the crosses through for more than two generations. But also unlike Mendel, he used many different species of plants—twenty in all. By the time he had finished, he was confident that particulate inheritance was general rather than some odd quirk of pea plants, a fact that Mendel could never have been sure of.

Then, quite by accident, he discovered Mendel's paper in an old file of reprints and realized that his work had been anticipated by thirty years. The paper that he then published, and the two papers published at the same time by other European plant breeders who had independently come across the same phenomenon, all gave priority to Mendel. The publication of these three papers in 1900 marks the real beginning of genetics as an experimental science. It is a small irony of history that if Mendel's paper had remained unnoticed in the dusty archives of the Verhandlungen naturforschung Verein in Brünn, we might now refer to de Vriesian genetics.

De Vries's youthful training in the observation of plants paid off in other ways, for he had already found exceptions to the Mendelian pattern. One plant that did not fit was the morning glory. Even before he began his systematic plant-breeding experiments, he had discovered some very odd-looking specimens of this flower growing along with the common variety in a weedy field in a town near Amsterdam.

When he planted their seeds, unexpected things happened. Some bred true, but many produced a mixture of types ranging from the common morning glory variety through others that did not resemble either their parent or the common type. It was as if these plants suddenly, according to laws that he did not understand, altered their genes dramatically to produce entirely different types. He gave these alterations the name *mutations*.

Mutations had been known for centuries, under the name *sports*. The chondrodystrophic mutants in dogs that had given rise to the dachshunds and other short-legged breeds, we now know were the result of true mutations. Plant and animal breeders already knew that some of them (the dominant ones, we now realize) bred true. But these mutations discovered by de Vries were something quite different. They often changed a whole suite of characters—size, shape and color of leaves, shape of stalks, and so on. Some had truly dramatic effects. One type, for instance, produced flowers with only female parts. And the rate at which these mutations occurred was remarkable—orders of magnitude greater than the rate at which sports appeared among domesticated animals or crop plants. Sometimes 50 percent of the progeny of two different parents would consist of one or more startlingly different types.

De Vries realized that he had found something very unusual. He used the results of these crossing experiments to construct a new theory of evolution. Microevolutionary changes, he suggested, were indeed due to the Darwinian process of natural selection. But larger changes were due to massive mutations, which occurred in a paroxysm of violent events. He saw little if any role for natural selection in these large mutations, which were so frequent when they occurred that they drove evolution willy-nilly.

Why did not all species undergo these kinds of violent changes? After all, if we were like morning glories, we would never be able to predict how many heads our children would have. De Vries assumed that not all species were "ripe" for these mutational events, that they had to accumulate this capability over long periods of time. But sooner or later, all would explode into fits of evolving.

This mutation theory of evolution gained many adherents among geneticists, who were still dazzled by the insight their new science had given into the way characteristics were inherited. If the genes controlled everything, then what need was there for natural selection? It was even suggested that there were species-level mutations that separated species, genus-level mutations that separated genera, and so on.

We now know that de Vries was not correct in his interpretation of his crosses. But it took almost thirty years, years of the most frustrating labor, before the mystery of the mutating morning glories was finally cleared up. Most of these events were not mutations at all, but rather the result of breakups through recombination of blocks of chromosomes that were normally passed as a unit from one generation to the next.

It was ultimately possible to explain all of de Vries's macromutations on the basis of known genetic phenomena. His evolutionary theory has fallen by the wayside. But was it entirely wrong? Is it not conceivable that real mutations, as massive in their effects as the apparent mutations de Vries discovered, might sometimes occur and be selected for?

Such mutations were proposed by the German developmental geneticist Richard Goldschmidt, whose inflammatory quotation (dating from 1940) heads this chapter. Goldschmidt's work on the fruit fly Drosophila led him to question the theory of the gene as a small discrete unit of information. Such neat little units of information had indeed been found in the bread mold *Neurospora*, but when Goldschmidt looked for them in Drosophila, he did not find them. His work led him to believe that genes were large and intricate structures that interacted with each other in often complicated ways, and that acted quite differently when they were moved around the chromosomes. Since he knew nothing of DNA, he imagined that these structures were made out of proteins. Genes like that might easily be changed in massive and un-expected ways. He called the poor creatures that experienced such mutations *hopeful monsters,* and suggested that occasionally such monsters might survive and push evolution in a new and different direction.

We now know that quite massive mutations do happen, but they do not normally result in anything as dramatic as hopeful monsters. For instance, at a time when our remote ancestors were fish, a very large mutation occurred in the lineage that eventually led to the amphibians, reptiles, and mammals. This mutation was a sudden doubling of the number of chromosomes, so that our ancestors suddenly had twice as many chromosomes as their immediate forebears. We can still see traces of this chromosome doubling in the way that the genes on our chromosomes are arranged. We know when it occurred because some present-day fish show the chromosome doubling while others do not.

This was a dramatic mutation indeed. But it happened in a line of our fish ancestors. They were fish before it occurred, and they were rather similar-looking fish immediately after it occurred. Its impact on the appearance and behavior of the fish was probably relatively small. If it had not been, the first mutant would probably not have survived.

Hopeful monsters cannot ever be ruled out (and they probably occurred more often in the past; see chapter 8), but they must always have been much less common than the ordinary workaday mutations on which the majority of evolution depends.

Even these ordinary mutations, however, can be very complicated events. Some have such pronounced effects that they are almost in the ''hopeful monster'' class. Most have much smaller effects, but they can still play an important role in evolution. Many of these mutations are much more complicated than the simple amino acid substitutions that I mentioned at the beginning of the chapter.

WHERE DO MUTATIONS COME FROM?

In the early days of science fiction, it was assumed by writers and readers alike that one hair-raising aftermath of an atomic war would be a ghastly array of mutant human beings, lurching about the ruined landscape and producing generation after generation of deformed horrors. It was supposed that mutational changes caused by radiation would be drastic and immediate in their effects. To this day, the public believes that one of the consequences of the atomic bomb explosions at Hiroshima and Nagasaki was a wave of mutants among the children of the survivors. This is not correct.

The Atomic Bomb Casualty Commission (ABCC) was set up immediately after the war to determine the effects of the bombs on the survivors and their children. Originally funded by the United States, it has now become a joint Japanese-American effort. The original commission has been replaced by the Radiation Effects Research Foundation.

Effects on the survivors were indeed severe. Stillbirths were common among pregnant women who were exposed to the radiation. In addition to the immediate effects of radiation sickness, there were many long-term health problems, particularly a high incidence of leukemia and other cancers. Even forty years after the explosions, there was a higher proportion of broken chromosomes in the white blood cells of the survivors than in the cells of a matched group of the same age from other parts of Japan.

But these effects are phenotypic, direct effects on the bodies of the survivors. Such effects are termed *congenital* effects, and they often have no obvious genetic basis.* Has there been any true genetic damage?

There were three major studies done on children of people exposed to radiation. Because the studies looked for genetic rather than congenital effects of the radiation, they were confined to children who were conceived *after* the explosions. One study examined whether there was any untoward outcome of the pregnancies, a blanket term that covered everything from stillbirths through congenital defects to true genetic damage. The second measured the survival of live-born infants. The third, started a decade after the war when the technology became available, examined the children for any chromosomal abnormalities. Further studies are now being carried out at the molecular level.

The control group consisted of an equal number of children born to parents from Hiroshima and Nagasaki, who were sufficiently far from the blast that they did not receive appreciable doses of radiation.

All three studies showed that among the children there was a slight positive correlation of deleterious effects with radiation dose to the parents. But none of these associations of radiation with harmful effects were statistically significant. What this means is that the slight correlation seen could have been due to chance alone.

The data gathered by the ABCC yielded correlation coefficients of about 0.001, one-thousandth of the maximum possible. The uncertainty with which this number was known was several times greater than the number itself, so that it was not possible to say that these coefficients differed significantly from zero.

The radiation effects were much smaller than some other effects that could be measured in the same studies. For ex-

---

* There may, however, be differing amounts of *susceptibility* among people to these environmental insults. That is, the genotypes of some embryos or fetuses may predispose them toward developing abnormalities if they are exposed to radiation at a critical developmental stage.

ample, it was also possible to measure the harmful effect of consanguinity of the parents on the outcome of pregnancy. This was found to be five times as great as the effect of radiation, and indeed the inbreeding effect was statistically significant.

Does this mean that there had been no genetic damage from the radiation the parents received? Most certainly not. But the damage was very slight, and indeed the data could be used to demonstrate that the human genome is very resistant to radiation. Using the calculated doses of radiation, and taking the tiny correlations obtained at face value, it was possible to determine that humans are at least four times as resistant as mice to genetic damage from a given dose of radiation.

This finding seems very strange. It has been known since the 1920s that X-rays and other high-energy radiation cause a great assortment of inherited mutations in all the experimental organisms that have been tested. What happened to all the mutations that must have been induced in the Hiroshima and Nagasaki survivors?

The answer, it appears, is that many of them were removed from the population very quickly by the action of natural selection. High-energy radiation tends to produce mutations with drastic effect, often breaking chromosomes and deleting pieces of DNA. An X-ray or a radioactive decay particle passing through a cell leaves behind it a string of charged molecules that can take part in a great variety of chemical reactions. As a result, a cell may suffer multiple "hits" from one ray or particle. If two of these hits happen to cause breaks in a chromosome, the deletion of a large piece of DNA can result.

Only damage to the germ cells that give rise to the gametes can be passed on to the next generation. (And not all of that is either; most sperm and eggs, damaged or not, are lost along with the genes that they carry.) A good deal of the germ-cell damage to the victims of Hiroshima and Nagasaki was drastic in its effects and expressed immediately. The result was spontaneous abortions so early in pregnancy that a woman may not even have known she was pregnant. In mice, even in the absence of external mutagenic agents, about 20 percent of fertilized embryos are lost in this way.

So far as the smaller mutations are concerned, there are efficient repair mechanisms in the nucleus of the cell that can actually heal the damage to the DNA caused by the radiation. Our repair systems are more efficient than those of mice, making us more resistant to the less drastic effects of radiation.

Indeed, we can even become more efficient at repairing our DNA. It is still fashionable in some European health spas to bathe in underground hot springs in which the water has high levels of natural radioactivity. The DNA repair systems of people who worked around these hot springs were measured and found to be significantly more active than the systems in people who worked elsewhere.

Our germ-line cells are not essential to the daily operation of our bodies, so that even if most of those cells die we can still survive. A fashionable method of contraception in the 1930s was a high dose of X-rays delivered to the male gonads; a man so treated would be unable to have children for several months but would be otherwise unaffected (at least in the short term).

In the case of whole-body radiation, damage to cells other than germ-line cells is more critical to immediate survival. The actual life-threatening damage occurs to rapidly dividing tissues such as those that produce the red and white blood cells. These are the first to be killed, and our lives depend on them.

The consequences of this sensitivity of critical cells to radiation became apparent during the ABCC studies. In order to produce statistically detectable numbers of harmful genes that could be passed on to the next generation, an enormous amount of radiation would have to be given. But as a result, the recipient would be likely to die from radiation poisoning before he or she had a chance to reproduce!

This is not the case with the fruit fly Drosophila. Insects are much more resistant to the phenotypic effects of radiation than are mammals; because there are few rapidly dividing cells in adult insects, their continued survival does not depend on them. Flies can absorb immense amounts of radiation, enough to kill a human ten times over, and still survive to produce offspring.

In the 1950s there was intense interest in the effects of radiation, since the planet was being showered with radioisotopes from atomic tests carried out by the United States and its allies, and later by the Soviets. Two major experiments were carried out to test the effects of radiation on Drosophila populations, one by Bruce Wallace in America and one by Gerd Bonnier in Sweden.

Over a period of several generations, the flies were given many times the lethal dose of radiation for humans. Bonnier's results were particularly striking. By the end of his experiment, the flies had gone through nineteen generations and had received thirty-eight thousand roentgens of radiation during that time (five hundred roentgens is a lethal dose for a human). Bonnier collected a fixed number of eggs each generation and found that at first many did not hatch. As a result, his populations dipped to about a third of his starting number.

But then they rapidly recovered. By the end of the experiment, they were back to the numbers at the beginning, *even though the immense blasts of radiation were continued unabated*. It was actually possible to select for flies that were radiation resistant. And these flies were quite capable of producing what appeared to be perfectly normal offspring. The science fiction nightmare of a wildly mutating population had not come true even under these extreme conditions.

X-rays and the highly energetic particles emitted by radioactive elements have proved to be tremendously useful mutagenic (mutation generating) agents and have been exploited extensively by geneticists. These agents are handy for breaking chromosomes. And because they produce multiple hits the chromosomes often rejoin in new and interesting ways. But this method of making mutations is crude, like hitting a watch with a hammer. The changes produced by radiation are mostly so drastic as to be eliminated quickly. The smaller changes that are made are also mostly harmful.

It has been suggested by some astronomers and paleontologists that the explosive bursts of evolution that are sometimes seen in the fossil record might have been due to a nearby supernova bathing the earth with high-energy radiation. This seems very unlikely. In general, radiation does not produce the kinds of changes that lead to adaptive evolution. To find the source of these changes, we must look elsewhere. Indeed, radiation has actually fallen rather out of favor among geneticists as a tool for making mutations. Much more sophisticated methods of mutagenesis have now greatly expanded the experimental possibilities.

The first of these was chemical mutagenesis. It was found in the 1940s that a great variety of chemicals produce mutations. Many of these chemicals act directly on the DNA, rather than charging through the entire cell like a bull in a china shop. As a result of this specificity, the spectrum of effects produced by chemicals tends to include a larger proportion of changes with small effect than does that produced by radiation. Ironically, and again because of this specificity, mutations produced by both man-made and natural chemicals in our environment may pose a greater danger to unborn generations than those produced by radiation.

The mutations produced by these chemicals were found to be simple breaks in the DNA, or small substitutions in the sequence of the bases that make up the genetic code. These changes tended to follow straightforward chemical laws. If the temperature was raised, for example, more mutants were produced because the chemical could interact more rapidly with the DNA. If more of the mutation-inducing chemical was added, again more mutants appeared.

But by far the most interesting mutagenic agents are those that do not follow these simple physical or chemical laws. Some of these agents have, quite literally, a life of their own.

INTRINSIC MUTATIONAL ELEMENTS

The most remarkable kinds of mutations are not produced by such *extrinsic* factors as radiation and chemicals, but by factors that are *intrinsic* to the cells themselves. The first of these were discovered in corn over forty years ago, by the redoubtable geneticist Barbara McClintock, who in 1983 finally received the Nobel Prize for her work.

McClintock was the beneficiary of two happy circumstances. She was one of the founders of a school of American geneticists who developed corn into a superb experimental organism. And she had at her disposal many mutants that had been selected, not only by her scientific forebears, but also by generations of North American Indians. The colored and spotted kernels of the Indian corn that makes its appearance every Thanksgiving on doors across the nation provided a wealth of genes for her to examine.

Corn is a hard organism to work with. Since the experimenter can raise only one generation a year, he or she must make every possible cross that might yield useful information and also try to anticipate every genetic eventuality. Unanticipated results always occur, so that crosses—usually the most interesting ones—fail or are not carried out and must be done the following year. A year is a long time to wait. And the crosses have to be done in the field, normally in the boiling sun.

In spite of these difficulties, McClintock observed that some of these genes that produced variegation did so by actually moving to other places in the genome and making all kinds of changes wherever they settled down. The movement of these genes was in turn the result of the effects of still other genes, which activated them.

The upshot of all these unexpected events was a dramatic alteration in the mutation rate in parts of the plant. Since it had been one of the fundamental tenets of genetics that genes did not jump about, her findings were startling to say the least. The work was widely discussed, but the crosses were complex and McClintock's reasoning was often difficult to follow. There was a feeling that there was something crazy about the phenomenon.

I recall talking about her papers with one of my professors in the early 1960s. He dismissed them with the explanation that chromosome doubling events in corn's past (like those that happened in our fish ancestors) had produced chromosomes that tended to adhere to each other in the wrong places and break and rejoin, giving the impression that the genes had moved about. Others had suggested that these genes jumped because Indian corn had been selected for so long for this kind of event. Callow youth that I was, these explanations satisfied me.

In short, her results were fascinating, complex, weird, and some sort of special case.

But in retrospect, they should not have been so startling. It had been known for decades that one of the hallmarks of evolution is a continuous rearrangement of the genetic material inside the chromosomes. This could be followed in detail using the giant chromosomes of Drosophila. Even if two species of fly looked superficially identical, when their chromosomes were examined they were often found to differ by dozens or hundreds of genetic rearrangements.

Now, unless all this genetic scrambling took place at some restricted time in the life of the species (shades of de Vries and his mutation theory!), there had to be some mechanism or mechanisms operating all the time to move genes around. In the clear light of hindsight, all this seems obvious. But it took decades for the significance of McClintock's results to be appreciated; that happened only when jumping genes were also found in bacteria. As a result, the field of jumping genes exploded.

You have to be tough to work on corn, but anybody can work on bacteria. These tiny cells have a generation time of as little as twenty minutes. Even though it takes many generations to produce a bacterial colony large enough to see on a petri dish, the experimenter can sit in air-conditioned comfort, seed single cells on a plate one evening, and have fine, large colonies to work with the next morning. Because each experiment takes a few days rather than a year, it has been possible to analyze the jumping genes of bacteria in unprecedented detail.

Some of the simplest of these are called insertion sequence elements (IS elements for short). IS elements are short pieces of DNA, sometimes as little as a few hundred bases in length, that have the capacity to make copies of themselves. These copies are inserted into other parts of the genome of the bacteria, and since they land fairly randomly, they can do a great deal of damage. If the copies are inserted into the middle of a functioning gene, they may effectively destroy the gene, because it will no longer be able to make a meaningful messenger RNA. Or, the copies may land in the middle of a critical regulatory region, where they may inappropriately turn a gene on or off.

About 1.5 percent of the genome of the common bacterium *Escherichia coli*, a denizen of our gut, is made up of these IS elements, so they are very important sources of genetic change for this species. Similar elements are also found in the genomes of higher organisms, and they can be even more plentiful. About 15 percent of the genome of *Drosophila melanogaster* is made up of elements that, like IS, have the capacity to insert copies of themselves elsewhere.

Computer programmers have recently and quite independently invented an analogue of the IS elements, in the form of "core wars" programs, or computer viruses. These terrifying programs were initially a form of entertainment, but it was quickly found that they can do immense damage to other programs and data bases in computers if they are let loose. Core wars programs come in a variety of types, but one simple version, directly analogous to an IS element, will generate a piece of nonsense and insert it at some random point elsewhere in the computer's memory. The program can be self-activating or can be activated by some inadvertant trigger in the regular program that happens to be running. Further, it is easy to program it to make copies of itself periodically, as insurance against accidental self-destruction or destruction by other core wars programs inhabiting the same computer.

Even a low level of activity of a core wars program is enough to bring any computer to a grinding halt. Unchecked, the core wars program will kill the computer.

Just as core wars programs are parasites of computers, IS elements act as parasites of the bacterial genome. And, as with other parasites, if they kill their hosts too readily, variants will soon be selected for that lessen their virulence. One sign of this is that most IS elements jump very seldom. On average, a given element will duplicate itself and send the duplicate to another part of the genome once in every ten million generations. This rate of jumping is equivalent to the mutation rate from other sources, one that the bacterial population can easily survive.

But there is another dimension to this complex interaction, which concerns the victim of this parasitic attack. The host can evolve the ability to fight back. The closest computer analogy to this is provided by the results of the first core wars tournament, which was recently fought out at the Computer Museum in Boston. The winners were two programs, inhabiting the same computer, that fought to a draw. One program made very large numbers of copies of itself, and the other lobbed pieces of nonsense into random points of the memory, destroying those copies as quickly as they were made.

In the living world it is possible for the host genome to fight back at a number of levels: by restricting the sites at which an IS element can insert itself, by making the cellular environment less favorable for IS element replication, and so on. We will see later that jumping genes are indeed distributed quite nonrandomly in the genomes of their hosts, something that would be expected if the host were fighting back.

It has recently been suggested that these IS elements could have a kind of life of their own, simply living in the chromosome and making more copies of themselves. This would be their only function. The term *selfish DNA* was suggested for these pieces of genetic material, though it might be easier (and less uncomfortably anthropomorphic) if we thought of them as pieces of parasitic DNA. The term *parasite* is, I think, closer to reality, for as IS elements multiply inside the genome they will inevitably cause more and more damage with their mindless leapings and insertions. It is the loss of susceptible hosts due to this damage, along with the evolution of the IS elements themselves, that probably regulates the numbers that they eventually reach on the chromosome.

BACTERIOPHAGE MU—AN EARLY VERSION OF AN IS ELEMENT?

In addition to the simple IS elements, there are more complicated infective agents that can produce mutations in bacteria. Some of these are viruses. There are of course many types of viruses that infect organisms of every level of complexity, but some of the most complicated viruses attack some of the simplest cells, the bacteria. These viruses are called bacteriophages (or phages), a name meaning "bacteria eater."

Bacteriophage Mu was the first real jumping gene to be found infecting bacterial populations. It was discovered in 1963 by Austin Taylor of the Brookhaven National Laboratory. He found that *E. coli* infected with this new bacteriophage showed an inexplicably high mutation rate, yielding every imaginable kind of mutant. Later it was determined that when the DNA of Mu enters the bacterial cell, it immediately inserts itself into the host DNA. And it does so with cavalier abandon, for it can insert itself not only into the bacterial chromosome but also into any other pieces of DNA that happen to be in the bacterial cell. These might be other viruses or other small pieces of DNA that are unable to form virus particles but that nonetheless can carry genes from one bacterium to another.

When Mu is integrated into the bacterial chromosome, it becomes what is called a prophage. Like many other so-called temperate bacteriophages, it can lie doggo in the bacterial cells for generations. But because it can insert itself anywhere, often disrupting the genes of the host, about 3 percent of the hosts become visibly mutant. That is, they lack one or another function that the wild type host used to have before the arrival of Mu. This is an astonishing mutation rate, about ten thousand times greater than the normal mutation rate in bacteria.

But the mutation process does not end with the arrival of Mu. In about one in every ten thousand infected bacteria, the phage DNA suddenly begins to multiply inside the cell. It does so in an unusual way, making repeated copies of itself, and again these are inserted at random in the host chromosome. These copies are, after an hour or so, chopped out and packaged in proteins to make mature phages. The cell breaks open and releases fifty to one hundred new particles. If there are any sensitive cells nearby these can be infected in turn, producing a new wave of mutations.

In view of all this genetic damage, it is difficult to imagine a population of bacteria surviving an onslaught of infection by Mu. In fact, however, it does so very nicely because in most cells the mutations occur in parts of the DNA that do not matter, or happen to damage genes that are not essential to survival. The bacterial population is able to multiply at a normal rate in spite of Mu—a situation reminiscent of those abused populations of Drosophila that were viciously bombarded each generation by X-rays and yet survived quite well.

Mu is one of the most complex and active of the jumping genes that have been found so far in bacteria. The mutations caused by Mu come in a great variety, and include deletions and rearrangements of pieces of the host DNA, linkings of Mu with the DNA of other phages, and the joining of different pieces of host DNA together. Although Mu happens to insert itself into the genome in a different way from that employed by IS elements, it may be that these latter elements started out as bacteriophages, lost most of their genes, and settled down to a calmer life hidden in the genomes of their hosts.

THE EVOLUTION OF MUTAGENIC AGENTS

This brings us to the matter of how these intrinsic mutational agents might evolve over time. Particular properties of the ends of the Mu molecule, and specific enzymes that it makes and borrows from its host, are necessary for it to do its genetic damage. It would not take many mutations of Mu, particularly in these critical ends, to make its insertion much more specific and thereby to lessen its destructiveness.

This specificity of insertion varies dramatically from one type of bacteriophage to another. Most of those that insert are much choosier than Mu about where they plug themselves into the bacterial chromosome. Some habitually invade only a few places, and others confine themselves to only a single site. Most bacteriophages, therefore, do not have the mutation-generating capabilities of Mu.

IS elements are much less destructive than Mu. Here the difference is not so much in the specificity of insertion, for IS elements can put copies of themselves into many different sites. Rather it is primarily in the rate of insertion. Mu always inserts when it invades a new cell, causing genetic havoc. But you will recall that IS elements, which are already part of the bacterial genome, make and insert their copies at a very low rate, comparable to the rates at which ordinary mutations occur. Mu causes a mutation to occur in a particular gene in one cell in every hundred thousand, while the average IS element might only cause a similar mutation in one cell in ten million.

We can actually see how these differences in transposition rate can evolve, because it is possible to change them in the course of laboratory experiments. The clearest of these experiments involve transposons. Transposons are another class of jumping genes that have the additional property of being able to move big pieces of DNA around. They consist of stretches of DNA with an IS element at either end. The activities of one large transposon, Tn10, can be followed because of marker genes that are included in the stretch of DNA in the middle. Mutations have been found in the flanking IS elements of Tn10 that greatly increase or decrease the rate at which it jumps.

It is interesting that only one other bacteriophage with properties as hair-raisingly destructive as Mu has been found in spite of years of searching by bacterial geneticists. Most jumping genes are by contrast fairly choosy about where they jump, and they jump at a fairly low rate. These are the properties one would predict for jumping genes that have had some time to coevolve with their host. It may be that in the natural populations in which it is found, Mu will gradually coevolve with its host and become less genetically antisocial with time. This can happen in two ways: either it will become less cavalier about where it enters the genome, or it may lose parts of itself and settle down to life in the genome, just as the IS elements have done.

Of course, even if Mu coevolves and loses its highly mutagenic properties, new Mu-like jumping genes will continue to appear. Natural selection will quickly determine whether they go on making mutations at a high rate or whether they will, with time, become more IS-like.

We have excellent evidence that the rate of mutation can change dramatically over evolutionary time. You will remember that one of the things that emerged from the atomic bomb data was that the genes of humans are several times more resistant to radiation than those of mice. It is a good thing for us that this is true, because if we were as susceptible to mutagenic agents as such short-lived creatures are, we would expect over the course of our much longer lives to accumulate dangerous numbers of mutations.

This resistance of our genes to change is also apparent from studies in which the rates of spontaneous mutation are measured. When these rates were measured at several gene loci in mice and humans, they were found to be about the same *per generation*. But because a human generation is one hundred times as long as a mouse generation, our rate of mutation per unit of *absolute* time is about a hundredfold less. Our DNA is therefore many times more resistant to change than is that of a mouse. This applies not only to radiation damage, but to the effects of other mutagens as well, presumably including transposable elements. It was essential to our survival as our life-spans lengthened that our DNA become more resistant to change.

But the most exciting implication of the discovery of transposable elements is that not only the rates but the *kinds* of mutations caused by them are under evolutionary control. Transposable elements are highly complex factors, and as a result the genes and the parts of genes that they affect as well as their jumping rates can all be modified.

The discovery of these elements has opened up a whole new, virtually unexplored field of evolutionary studies and has dramatically changed the way we look at the mutational process.

JUMPING GENES IN OTHER ORGANISMS

How important are jumping genes in more complex organisms, or are bacteria somehow unique?

Bacteria have pointed the way, so it is now possible to investigate in detail the molecular effects of jumping genes in higher organisms. In later chapters we will look at many examples. Jumping genes have now been found to be important in insect development, in the rapid adaptation of parasites to their hosts, and even in the generation of genetic variability within a population. Indeed, the rules for jumping genes in higher organisms are just beginning to be explored. Let me tell about one case in which our group was peripherally involved.

You will recall from the beginning of this chapter that when yeast are denied the use of oxygen, they will die unless they have the alcohol dehydrogenase enzyme. We selected mutants of this enzyme by denying yeast oxygen and then giving them a poisonous alcohol or an inhibitor of the enzyme. The cells that survived made mutant enzymes with altered function.

Other workers turned this process around by taking mutants that already lacked this enzyme and denying them oxygen. The only cells that could survive were those in which the alcohol dehydrogenase gene had been suddenly turned on.

They found that many of the mutants in which this happened were the result of the insertion of IS-like elements in front of the alcohol dehydrogenase gene, where there was a control region. The original mutant lacked the enzyme because this control region had been modified to turn the gene off permanently. The arrival of the new copy of the IS element from elsewhere in the genome disrupted the control region, with the result that now the gene was turned *on* permanently.

Two of these workers, Charlotte Paquin and Valerie Williamson, designed an experiment to determine the conditions under which the jumping gene could be persuaded to insert itself most readily. They found, for example, that lowering rather than raising the temperature increased the rate at which the genes jumped. This is quite the reverse of the usual situation seen with ordinary mutagenic agents, in which the rate of mutation increases with increasing temperature.

They showed that the rate of mutation due to IS jumping increased at least one hundredfold as the temperature dropped from twenty-five degrees centigrade, a little above room temperature, to fifteen degrees, the temperature of a sunny fall day with a little nip in the air.

Interestingly, the activity of jumping genes in bacteria and in plants has also been shown to be greatly stimulated by low temperatures. One is tempted to surmise that these genes jump to keep warm, but there is a more prosaic explanation. It has been shown in a bacterial system that one of the enzymes responsible for the transposition process is temperature sensitive, so that it is inactivated at high temperatures.

This makes an excellent proximal explanation. But it begs the question of why the jumping genes in organisms as diverse as bacteria and plants all show the same pattern of temperature sensitivity. This property must somehow have been selected for, even though we cannot at the moment see why such a temperature sensitivity might have evolved. Nonetheless, this strong dependence on the environment illustrates the complexity of the IS insertion process and shows how natural selection must have shaped it.

We will also see later how factors other than temperature can actually stimulate the production of remarkably specific mutations, some of them apparently connected to the factors that produce them. Lamarckism will rear its ugly head again, and we will do our best to lop it off.

It is strange, and a little unnerving, to think that there are parasites living within our cells that are not only capable of changing our genes, but also capable of altering the ways in which they do it. These are not separate organisms. They depend on us utterly for their own survival. But they are capable of doing us great harm. Just as bacteria and viruses attack our bodies, forcing us to evolve an elaborate immune system to deal with these invaders, so jumping genes and integrating viruses attack the store of information that makes us what we are. We must in response have evolved mechanisms to defend ourselves against them, though we have as yet few hints of the shapes these mechanisms may have taken.

We have proof that these genetic battles have taken place because some of the parasites of our DNA are highly sophisticated. Mu is a blunderer by comparison with the mammalian retroviruses and the related pieces of DNA that have recently been found in yeast and Drosophila. Over time, many of these parasites have become expert at lurking in our genes, doing harm only occasionally and perhaps even sometimes benefiting us. This kind of accommodation is important for any parasite, but it is absolutely essential for the long-term survival of parasites that live inside the nucleus of the cell. This is because many of them are trapped there and are quite unable to escape. Their fate is the fate of the host. Nuclear parasites that damage the host too much bring certain doom upon themselves as well.

Not only have we evolved mechanisms for diminishing the damage these parasites do to our genes, but it seems likely that with the passage of time the very structures of many of our genes have coevolved with our mutagenic agents in a mutually beneficial fashion. We will return to this theme many times as we look at some of the new information that is emerging about our genetic architecture. Such a progression from host versus parasite to mutual benefit probably lies behind many of the cosy relationships we have formed with other organisms. Even structures within the cell, like the mitochondria on which we depend for much of our energy, might originally have started out as harmful parasites.

This continuous genetic warfare, shading into temporary truces and alliances, must have had a large hand in shaping our own genes. Science fiction holds nothing stranger than this drawn-out, invisible battle between ourselves and these parasites on the very stuff of our being.

5

Of Tuxedos and Antibodies

One other thing will the professed mathematicians say about this thoroughly bad and vicious book: that the reason why it is *so easy* is because the author has left out all the things that are really difficult. And the ghastly fact about this accusation is that—*it is true!*

—SILVANUS P. THOMPSON, *Calculus Made Easy*

THE TOOLS OF THE MOLECULAR TRADE

Some years ago I spent the whole of one long, dark winter in the city of Stockholm, taking a protein apart bit by bit. There was little else to do. Stockholm is wonderfully organized for winter sports, so it was possible to take breaks and go skating on the lakes or cross-country skiing. But the dollar was weak, and a trip to a restaurant or the theater was quite beyond our budget (and beyond that of most Swedes as well). By the end of the winter we understood the saying that in Sweden the standard of living is so high nobody can afford it.

In spite of all the time I spent in the lab, it took the whole winter to learn about two-thirds of the amino-acid sequence of the protein I was working on. Although the techniques I used were the state of the art at the time, they have since been largely displaced by new, much simpler, and more powerful methods. As a result, knowledge about proteins and the genes that code for them is now expanding so rapidly that one of the biggest problems is how to organize this new information so that researchers can get at it.

Armed with this ever-growing collection of tools, molecular biologists are beginning to explore the genetic material of many different organisms with the zeal of kids let loose in a candy store. And they are finding remarkable things, not only about the genes themselves, but also about the ways that they are organized.

It is this organization that interests us here. We would like to see if genes are organized in a way that facilitates future evolution. Put another way: Has it become easier for a typical population of organisms to respond to a new environmental pressure than it was in the past? The degree to which the genes are organized should tell us this.

A homely analogy will make this clear. The office in which I am working is a wilderness of books and papers, strewn higgledy-piggledy. There are days when I spend more time looking for things than getting things done. Some of my friends, by contrast, have beautifully organized offices. They undoubtedly get much more work done than I do. (Perhaps the work is not as original and insightful, but the quantity is greater.)

So it is with the genes. The very fact that a genome is well organized allows adaptive changes to take place in it more easily than if the genes for various functions were scattered about with no particular relationship to each other.

This chapter and the next will be spent examining this possibility. I want to introduce you to a number of protein molecules that show such an organization and that as a result are able to evolve with ease to meet new situations. In the case of the protein dealt with in this chapter, it is clear that selection acting on individuals had the most to do with shaping this organization. The story is less clear for the protein family that will be explored in the next chapter, since the selection that shaped those genes occurred in the very distant past. It may be that species selection played an important role in those distant events. But the result in both cases is that the genes are now molded in such a way that further alterations leading to adaptive changes can occur easily. These genes can now play an important role in the facilitation of further evolution.

Let me begin with the immunoglobulin molecule.

THE IMMUNE SYSTEM AND HOW IT WORKS

The immune system presents one of the most complex and scientifically challenging problems in all biology. We have discovered how it protects us against diseases that have preyed on our species for thousands or millions of years. More remarkably, we have found that it can protect us against diseases that we have not yet met. And it does so by means of an array of proteins, the immunoglobulins, that can bind specifically to molecules that they also have never met!

This begins to sound like the very thing we want to shy away from as we talk about evolution. How is it possible for our immune system to see into the future and construct immunoglobulins that help us attack new diseases? Of course, our immune system does not see into the future. But it *appears* as if it does. It turns out that this unnerving appearance of precognition on the part of the immune system tells us a lot about evolutionary facilitation and how it might work.

The immune system, as we will see, is highly efficient. But it can be pushed beyond its limits. Three or four generations ago, bacterial infections were an ever-present and terrifying factor in human affairs. Typhoid, typhus, plague, cholera, and tuberculosis were major causes of death (as indeed many still are in the third world). In England in 1850, thirty children in every hundred died before puberty, usually in infancy.

Even leaving aside infant mortality, the average lifespan was only forty years. And this was in spite of our immune system, developed over hundreds of millions of years of evolution, which ought to protect us against such diseases. It does, but the life-style of the time had largely defeated it.

Victorian London was in worse shape than most third world cities today. Whenever it rained hard and the sewers leading into the Thames overflowed, Parliament had to adjourn because of the stench. Henry Mayhew, the pioneering English sociologist, tells of visiting London tenements in the 1850s, to find that decades of human excreta had been allowed to pile up in the courtyards. This was because thoroughly aged night soil was useful in the tanning industry.

Just a few years before Mayhew wrote his pathbreaking study, *London Labour and the London Poor* (published between 1851 and 1862), the collecting of night soil in city courtyards had been outlawed. But it was still carried on clandestinely in the heart of slum areas like the tenements around Rosemary Lane, where the authorities never ventured.* In other parts of the city, ''sewer finders'' plied their trade in the city sewers, looking for items of value among the offal and fighting off swarms of rats. We have Mayhew to thank for documenting the city's filth and poverty, but London was certainly not unusual among large cities of the time.

Then things began to change. On the Continent, Louis Pasteur and Robert Koch demonstrated that bacteria cause disease. They had to fight the medical establishment, heirs to centuries of prejudice and stupidity, but gradually their view began to prevail.

The Hungarian physician Ignaz Semmelweis, working in

---

* Rosemary Lane leads from the Tower of London east toward Cable Street, and when Mayhew knew it, much of it was occupied by a vast tobacco warehouse in the midst of the teeming tenements. It has changed dramatically since that time. The warehouse has been replaced by the Royal Mint, the tenements have been replaced by banks and upscale pubs, and Rosemary Lane has been totally transformed into Royal Mint Street. From filth to filthy lucre in a few generations!

Vienna, showed that something as simple as doctors washing their hands before examining the next patient could greatly reduce the spread of infectious disease. Semmelweis was treated dreadfully by the medical establishment for suggesting that cleanliness might save lives; he was driven to insanity and suicide. Eventually, however, doctors did learn to wash their hands, and they now make quite a ritual of it.

But even before these discoveries, the rise of an affluent and educated middle class, which did not want to be exposed to filth and stench, brought about a demand for public health measures. And changes in the way people lived began to reduce the incidence of infectious disease in the West well before the introduction of antibiotics in the 1930s and their widespread use in the 1940s. It seems that the immune systems of people living in the nineteenth century and earlier were simply overwhelmed by massive bacterial infections, the result of ignorance, poverty, and crowding. Reducing the dirt in the environment allowed the system to work properly.

How does this remarkable system protect us? Its primary function is to produce proteins called *immunoglobulins*, which are also known as *antibodies*. These bind to foreign proteins or carbohydrates that enter the body, either as single molecules or on the surface of invading cells. These foreign materials are called *antigens*, because they can generate an *antibody response*. Once bound, the invading proteins or cells are inactivated, and other parts of the immune system are called into play to destroy them.

The system is enormously complex, and much of this complexity has to do with the fact that invading cells, viruses, and foreign matter can arrive in the body in a great variety of guises. The immune system must be able to deal with all of them.

In mammals, the system has its start in the liver and later in the bone marrow of the developing fetus. Many types of cells begin to differentiate there. They all have the same genetic information initially, but over time they begin to exhibit very different behaviors and abilities. And as they develop, some of them even begin to rearrange their genetic information.

The various types of cells that make up the immune system all begin as undifferentiated stem cells. By the time they have finished developing, they can perform a variety of specialized tasks. And they interact with each other in tangled ways that are not yet fully understood.

An invading foreign organism or material is most likely to meet a macrophage first. These most voracious of the white blood cells circulate throughout the body. Macrophages are perhaps the most primitive type of cell in the immune system. Immunity-like reactions involving such cells have been seen in the insects and their relatives, and indeed the first observation of white cells engulfing invading organisms was made by the Russian scientist Elie Metchnikoff at the end of the last century. He saw what he called *phagocytes*—''cell eaters''—destroying foreign particles in the transparent bodies of water fleas.

If the body has never experienced the invader before, only a few antibody molecules will recognize it and will bind to it. This is usually enough to stimulate macrophages to engulf the foreign material.

Macrophages are rather messy eaters. Bits of the foreign matter cling to the outside of the macrophages, binding strongly to molecules somewhat related to antibodies that are floating in the cell membrane. This in turn allows the macrophages to interact with the next major class of cells in the system, the T cells.

T cells mature in the thymus, an organ that functions only in childhood and then disappears. Mature T cells are covered with receptors that bear a stronger resemblance to antibodies than do the molecules on the surface of the macrophages. T cells carrying receptors that recognize the foreign antigens are stimulated to multiply through the action of yet other molecules called lymphokines, which are released by the macrophages.

Many types of T cell result. The first of these are helper T cells, which interact with other categories of T cells and turn them into killer cells that can detect antigenic proteins that have been left behind on the surface of infected body cells. The killer cells then kill the infected cells immediately. Perhaps most importantly, helper T cells also interact with yet another category of cell called B cells and stimulate them to make antibodies against the foreign antigen.

The antibody is the final line of defense in this complicated system. It is fascinating to see how the specificity of the system increases with each stage in the process. First, the antigen binds, though rather unspecifically, to the molecules on the surface of the macrophages. This in turn enables these complexes to interact with the receptors on the surface of those T cells that are most specific for them. T cell receptors come in a greater variety of types than the molecules on the macrophages, but they are still less variable than true antibodies. The T cells that best match the macrophage complexes are only a tiny fraction of all the T cells in our system, but they multiply quickly once they are stimulated.

Then the T cell receptors interact with antibodies on the surface of a tiny fraction of the B cell population. These B cells are then stimulated in turn to make the free-floating immunoglobulin antibodies that patrol all the parts of our body in the blood and the lymphatic system. As a result of this interaction with T cells, this subcategory of B cells begins to multiply and within a few weeks makes more and more of these particular antibodies, which are highly specific to the invading antigen.

The antibodies can coat the invaders and make them more toothsome for macrophages. They can also activate yet another line of defense called the complement system, which punches holes in invading cells. But the increase in specificity of the response does not stop here, as we will see. The B cells continue to refine their antibodies, making them a better and better fit to the antigens. The marvelous success of the immune system lies in this cascade of interactions that make growing amounts of increasingly specific molecules for the body's defense.

During the early differentiation of the immune system we learn not to make antibodies to the proteins in our own bodies. But this self-recognition process occasionally fails, and the resulting autoimmune response can lead to very serious problems. Among other diseases, many types of arthritis are now suspected of having an autoimmune component.

Even for those of us who live in temperate regions, bacterial and viral infections are a continuing danger. They enter our bodies in our food, through wounds and sexual contact, and on dust and in tiny aerosol droplets from the breath of others. Most are destroyed at the site of entry, either through the action of digestive enzymes or by the normal activity of the immune system. Small numbers of any new bacterium or virus can usually be dealt with because we normally make antibodies that bind, though not very well, to the antigens on the surface of these new invaders. But this first line of defense can be overwhelmed by a massive initial infection of the type that was common before the spread of public sanitation.

The great power of the immune system, however, lies in its ability to learn from experience. The first few invaders will stimulate the immune system to produce more of the B cells that make antibodies specific to them. Further, by several molecular mechanisms that are now understood in some detail, the new B cells now make better antibodies with a higher affinity for the invaders. This multiplication of cells, and the fine-tuning of the antibodies they make, is called a secondary immune response. If the same organisms invade our bodies again, even if they come in much greater numbers, they find our immune system ready to deal with them.

Sophisticated as this secondary response is, it cannot help

us unless our body is able to mount some kind of primary response to any kind of invader. To make this clear, let me make what seems at first sight to be an absurd comparison. Let me compare the immune system to a tuxedo rental store.

For such a store to be successful, a great variety of tuxedos must be available for potential customers, even though each individual tuxedo may be rented rather seldom. In southern California this can be carried to extremes. A former technician of mine was asked to be best man at a wedding of a friend who had some claim to Irish ancestry. My technician was required to find, at short notice, six Kelly green tuxedos with matching shoes in various sizes. When he told me of this task I dismissed it as impossible. But he was able to do it with a single phone call!

Our basic immune system must be similarly adaptive. Unstimulated B cells each produce one of a highly diverse collection of antibodies called *IgM*. Each of these is a cluster of five identical immunoglobulin molecules. In turn, each molecule has two sites that can bind to the antigen, so there are ten sites in all. Because of these multiple sites, even an IgM molecule that is a rather poor fit to the antigen can bind much more tightly than if it consisted of only a single immunoglobulin molecule.

We find the same thing with tuxedo rentals. Tuxedo stores cannot have every size, but an ill-fitting tuxedo is better than no tuxedo at all.

Once the basic system has been stimulated by an invader, the selection and fine-tuning processes produce antibodies that are more specific for the invading antigen. The molecules produced by this response, called *IgG*, are single rather than multiple. Because their affinity with the antigen is so great, they do not need to be bound together in clusters.

Even when it has never been challenged by antigens, the body produces a great many different kinds of IgM molecules. A mouse makes somewhere between one and ten million different types of these antibodies, even though it may have been raised in an environment made as free as possible of bacterial and viral infections.

The protein and carbohydrate antigens themselves are not infinite in their variety. If one puts the antibody-producing cells of a single mouse in a test tube, then introduces a new protein antigen to which the mouse has never been exposed, about one-hundredth of one percent of the cells (one cell in every ten thousand) bind the protein. Since the mouse makes at least a million different IgM antibodies, this means that at least a hundred of them can bind to some degree to this new antigen.

This experiment works regardless of what protein is introduced, no matter how foreign. Of course, a different subset of the antibodies binds to each new introduced antigen.

The system is sufficiently diverse that it can handle any combination of the twenty amino acids that make up proteins. Indeed, it even works if, as is often the case, some of the amino acids of the invading organism have been modified so that they are different from the twenty we normally find in proteins. Further, the system can recognize any combination of the dozens of sugars that make up carbohydrates, for these compounds are often present in the cell walls of bacteria or on the surface of viruses.

HOW THE IMMUNE SYSTEM GENERATES ITS FINAL LINE OF DEFENSE

If tuxedos were made in one piece, so that people had to climb in through a hole in the back, tuxedo rental stores would have to stock an even greater variety of these garments than they do. Luckily, it is possible to mix and match pants, coats, shirts, and shoes of different sizes. The same applies to the immune system. A simple but wasteful way to generate all the millions of antibodies in our final line of defense would be to have one gene in our genome for each type of antibody. But it would be much less wasteful to carry a fairly small number of parts of genes that can be mixed and matched to make the complete antibody. We now know that the evolution of the immune system has taken the second route. There are a couple of excellent reasons why this is so. I will deal with the first now and the second a little bit later.

The first reason has to do with the difficulties of evolving a "one gene, one antibody" system. Let us assume that there was, in the distant past, an ancestor of ours that carried a single antibody-producing gene. The antibody might be quite primitive, perhaps simply a protein that bound to the surface of some important parasite. This in turn might make the parasite more susceptible to digestion or perhaps more likely to be engulfed by phagocytes.

The parasite would have had an easy way to circumvent its host's defense, however. Almost immediately, a mutant strain of the parasite, to which the host's antibody protein could no longer bind, would be selected for. Now, some parasites would have the altered protein and some would have the original one. Parasites with the altered gene would have the advantage, and the mutant gene would quickly replace the old gene in the parasite population.

The next move in this evolutionary game might be a gene duplication in the host, so that it now had two antibody genes. This would free the new gene to alter over time and eventually produce an antibody against the parasite's new protein.

Now it would be the parasite's turn again. A new strain of parasite could easily arise. Further, new and different parasites could put in an appearance at any time, carrying proteins so alien that it would be very difficult for the organism to alter a duplicate of its antibody gene enough to recognize them. Just as it is easier to build more ICBMs than it is to find ways to destroy them, the parasites could always be a jump ahead of their hosts. The host species would have to wait, first for a gene duplication to occur, and then for the duplicate to be suitably modified.

Since it would all depend on chance mutational events, this might take a very long time, but it is not as unlikely as it might seem. When duplicated genes lie next to each other on the chromosome, the resulting situation is genetically unstable. In the case of long stretches of such genes, the process of genetic recombination works rather inaccurately because the wrong parts of the homologous chromosomes* tend to pair up. As a result, major duplications and deletions of the genetic material can be passed on to the next generation, through the process of *unequal crossing-over*. A simplified diagram of such an event is shown in figure 5.1.

Eventually, this evolutionary process would produce enough antibodies to be able to bind to any antigen likely to challenge the organism. One large problem with this hypothetical scheme is that each antibody would have to be individually shaped by selection. Another problem, which can be seen from the figure, is that deletions as well as duplications are produced through unequal crossing-over.

Deletions can occur in the tuxedo rental business, too. The new owner of a rental store might be inclined to throw out those tuxedos in odd sizes and outré colors that had not been rented for years. But if heliotrope or Kelly green tuxedos suddenly became all the rage, he could be in difficulty. Similarly, if the genes that were lost through a deletion happened to specify antibodies that were not needed at the time, the chromosome carrying the deletion might spread by chance through the population, and the deleted antibody genes might be lost permanently. The products of thousands or millions of years of painful evolution might be lost to the species through an accident.

---

* Homologous chromosomes are the pairs of chromosomes we inherit, one from each parent, that carry sets of genes that perform the same tasks. Genetic recombination, resulting in the exchange of parts of these chromosomes, occurs in the course of the formation of our gametes.

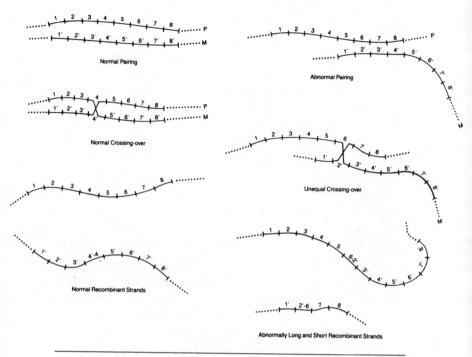

Figure 5.1 This diagram shows how the process of unequal crossing-over can generate duplicated copies of a gene. In this simplified picture, only two of the four strands involved in pairing are shown. One chromosomal strand is marked P to show that it was originally derived from the organism's father; the other is marked M to show that it was originally derived from the organism's mother. After recombination, which mixes the chromosomes up very thoroughly, this distinction is lost. In this particular case, genes for the ribosomal RNA of Drosophila pair up incorrectly in the course of production of the sex cells. In the example on the right, the process of recombination, which would ordinarily simply scramble the order of the genes, now produces one DNA strand with too many genes and one with too few. It is worth remembering that while the unequal crossing-over can result in the gain of genetic material, it is equally likely to result in its loss, since the shorter DNA strand has the same chance of being passed on to the next generation as the longer strand has.

The danger posed by deletions is to some degree removed because of the way the immune system did evolve: it did not require the serial evolution of such a large number of separate genes. Nonetheless, it has resulted in an immune system that can easily and quickly generate such a variety of different antibodies that the host can deal with any imaginable number of new and evolving parasites. The details of how this is done have been recently worked out using the techniques of molecular biology.

It was discovered quite early that certain cancers of the immune system result in an enormous overproduction of one type of immunoglobulin-producing cell, which in turn produces only one type of antibody. These cancers provided immunologists with an enormously potent tool, a *clone* of cells that made not only a single type of antibody but also a single type of messenger RNA for that antibody instead of the millions that are produced in an immune system that is functioning normally. Much more sophisticated ways are now available for making clones of particular antibody genes, but even before these were invented immunologists had harnessed cancers to produce usable clones.

It had long been suspected that the number of antibody genes is smaller than the number of antibodies. The first clearcut experiments to show this were carried out by using the purified clones of messenger RNA as a template for making radioactive DNA strands. These radioactive pieces were then hybridized to DNA taken from normal cells. Although various immunoglobulin genes do differ slightly from each other, they are sufficiently similar that such a radioactive probe made from one gene can be made to hybridize to similar genes elsewhere in the genome. The radioactivity that stuck to the DNA after hybridizing gave a measure of the number of copies of genes that were similar to the cloned gene. This turned out to be surprisingly small, in the hundreds rather than the millions.

More sophisticated experiments done later showed that these genes could actually move around the chromosomes as the antibodies were being assembled. It was found that if cloned DNA made from the various parts of an antibody gene was hybridized to DNA from cells that did not make antibodies, the different parts hybridized to different places in the genome. When the same experiment was done using DNA from cells that made the antibody, the parts all hybridized to the same piece of genomic DNA. They had somehow been moved adjacent to each other.

This was a startling finding, as astonishing in its way as the discovery of jumping genes. The immunoglobulin genes are put together from widely scattered parts, just as a rented tuxedo is assembled from various tuxedo parts.

We can now understand the second reason why the immune system has evolved in the fashion that it has. The difficulty we mentioned earlier, that some genes might be lost from the population if a deletion produced by unequal crossing-over spreads through it, becomes much less of a problem. This is not because the likelihood of loss is less when the genes are fragmented, but rather because the consequences of loss are much more severe to the individual. Each gene fragment takes part in the production of many different antibodies, so if it is accidentally deleted its loss is likely to be felt immediately. The organism unfortunate enough to lose some of its antibody gene fragments is at a much greater disadvantage than it would be if it lost a few complete but seldom-used antibody genes. Because of this immediate disadvantage, it is much less likely that a deletion of some of the antibody fragments would spread by chance through the whole population.

It is as if, in an excess of housecleaning zeal, the owner of a tuxedo rental store threw out all the pants. He would soon go out of business, regardless of how great a variety of shirts and jackets he had in stock.

MAKING AN ANTIBODY

I would like to delve into some of the detail of the complicated process of putting together a functioning immunoglobulin molecule in order to make some important points about
how the process has evolved.

Each immunoglobulin molecule consists of four chains of
amino acids. Two of these proteins, which are identical to
each other, are about twice the size of the other two, which
also form an identical pair. The large ones are called heavy
or H chains, because they can be separated from the smaller,
lighter L chains in a centrifuge. The H and L chains are attached to each other in such a way that one end of the complete molecule can bind either to other immunoglobulin molecules or to the surface of an immunoglobulin-producing cell.
The other end, however, is the real business end of the molecule: this is the end that binds to the invader's proteins or
carbohydrates.

The business end of both the H and L chains consists of a
variable, or V, region of about 110 amino acids. The rest of
the H or L molecules are much the same from one antibody
of a given type to another, and are called *constant*, or C,
regions. It is the variable regions, which differ dramatically
from one antibody to the next, that confer such awesome
variety on our collection of antibody molecules.

Putting together a single subunit of an immunoglobulin
molecule, either a light or a heavy chain, involves joining
three genes together: a V gene, a J (for joining) gene, and a
C gene. The probe experiments tell us how these genes are
arranged on the chromosomes.

The genes that will contribute to the L and H chains are
on separate chromosomes. On each chromosome the V
genes—as many as several hundred of them—come first. They
are by far the most numerous class of immunoglobulin genes,
though they are by no means as numerous as the millions of
genes that would be necessary if we had to have a separate

gene for each antibody. After the V genes there is a long stretch of DNA of uncertain function, followed by a cluster of a small number of J genes. Then, very near the J genes there is another small cluster of C genes.

In the construction of a mature antibody messenger RNA, a V gene, a J gene, and a C gene are put next to each other. This involves the removal of large intervening pieces of DNA or RNA: this removal takes place as each antibody-producing cell matures. The final assembled molecule specifies a particular L or H protein, and since this same process is carried out for both the L and the H subunits, the number of possible ways to make a complete antibody molecule is very large. Only one L-carrying and one H-carrying chromosome take part in this process, so each cell ends up making one antibody. Some of these complicated maneuvers are shown in figure 5.2.

The diversity of antibody types is increased yet further by other mechanisms. For instance, the V-J join is not precise, and can vary. In the construction of the final H-specifying molecule, another small segment (the D segment) is inserted into the V region.

Even after this assembly process, the antibody gene can continue to change during the life of the antibody-producing cell. During the period that the selected antibody-producing cells continue to proliferate, resulting in the secondary immune response, parts of the V region show a high rate of mutation. These mutations are called *somatic (body)* cell mutations because they do not take place in the germ cells and are therefore not passed on to the next generation. Cells carrying somatic mutations that permit the antibody to bind more strongly to its antigen will increase in number. Somatic mutation and somatic selection contribute to the fine-tuning of the match of antibodies to their antigens, which is seen in the secondary immune response.

Returning to our metaphor, it is worth noting that some of the classier establishments that rent evening wear, such as

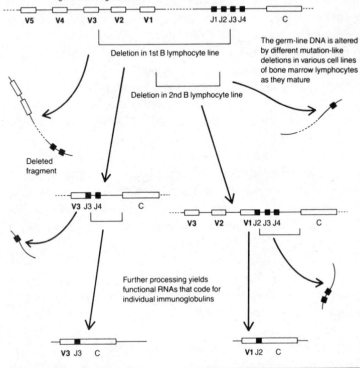

Figure 5.2 The process by which an immunoglobin light-chain gene is put together. A simplified version of the germ-line sequence is shown at the top of the figure, with some of the V, J, and C genes. The diagram shows two different sets of deletion events by which mature immunoglobulins are produced. Many other similar events in different cell lines produce the diversity of antibody genes from the pieces in the germ line. The exact sequence of events varies depending on the kind of chain being made. In this example from the mouse, some further messenger RNA processing is necessary to join the J and C genes and produce the mature message.

Moss Bros in London, have tailor shops on the premises to make last-minute alterations. There is a marvelous point-for-point resemblance between the immune system and the tuxedo rental business.

This whole remarkable genetic construct disappears when the cell dies. The piecing together of the antibody gene and the subsequent V-region mutations all take place in somatic cells, so that none of this new information is passed on to the next generation. But the basic collection of V, J, and C gene fragments in the genome *is* passed on. And these have the capability of making, each generation, a new and highly diverse collection of antibodies: subsequent somatic mutation and selection tailor the immune systems of each generation to the antigens that they actually meet.

As a consequence of gene fragmentation and gene duplication, each antibody-producing cell has the potential to make one of millions of different antibodies. Thus our immune system is ready to respond to anything that the natural world throws at us, so long as we are not already weakened by disease or our systems are not flooded by the new antigen.

PRESSURES THAT HAVE FORCED THE IMMUNE SYSTEM TO EVOLVE

The immune system is the product of hundreds of millions of years of evolution. It is present in its most elaborate form in the birds and mammals, but simpler immune systems are found in most of the vertebrates and even some other groups. You will recall that Metchnikoff saw in water fleas the first stages of an immune system, the activity of phagocytes.

Simple creatures such as water fleas do not show all the sophisticated elaborations of our own immune system, and because they live for such a short time and have such a high reproductive rate they may not need them. But, while the immune system of higher organisms has been growing in complexity in the course of evolution, our enemies have not stood still. We are now faced with an array of parasites that have found ingenious ways to get around our defenses.

For example, the pneumococcus which causes bacterial pneumonia coats itself with a polysaccharide, a sticky substance made up of polymerized sugars. The sticky coat has the effect of enormously increasing the surface area of each bacterium, so that much more antibody has to be produced than if the bacteria were naked and exposed. In the days before antibiotics, only about a third of the victims of bacterial pneumonia survived, and they did so because they were eventually able to produce so much antibody against the polysaccharide coat that the pneumococcus could be "seen" and destroyed.

There are over eighty different types of pneumococcus, each producing a slightly different polysaccharide coat. One of the reasons that some of these people were able to survive pneumonia infections while others were not was that their immune systems were able to recognize certain pneumococcal strains more effectively. As with all our other internal parasites, there is a race going on in which the ability of the parasites to generate more types is pitted against the ability of the immune system to recognize them. Not all immune systems are equally good at handling all the pneumococcal variants. That is why our defenses were defeated so often before the advent of antibiotics.

Viruses have found a different way to get around our defenses. They spend much of their time inside the cells of their hosts, where they are protected against the immune system. If, in the process of inserting themselves into the host cell, they leave bits of their outer coat embedded in the cell membrane, the immune system may detect these foreign proteins and destroy the cell before the viruses have a chance to multiply. If they do not, the viruses may be harbored safely in the cell for long periods. Herpes viruses may be concealed in cells for decades and only make an occasional appearance. There are many other so-called slow viruses as well. Adept at disguising themselves, they work their damage inside the cell over the course of years.

More complex parasites also have the ability to disguise themselves. One great difficulty in designing an antiserum against malaria is that the parasites, once introduced into the host, immediately migrate to the liver, giving little time for the host's immune system to react. When the parasites re-emerge from the liver cells and invade red blood cells, they have dressed themselves in a completely new coat of antigens, again catching the immune system by surprise.

The tiny worms of schistosomiasis, which can block the circulatory and lymphatic systems with their remains when they die, are even subtler. They exploit our immune system's self-nonself distinction to turn it against us. Once they enter the body, they quickly coat themselves with a layer of our own blood proteins, then slip through our circulation like the Scarlet Pimpernel through revolutionary France.

Because they go through many generations to our one, the parasites can evolve with blinding speed. Special mechanisms in many viruses ensure that the genes making coat proteins mutate at a higher rate than the rest of the genes. In the viruses that cause the common cold, it has recently been found that the parts of the coat proteins that stick out can change much more rapidly than the parts that are concealed in clefts on the surface. It is this high rate of change in specific genes and parts of genes that makes it impossible for the victim to produce large quantities of enough different antibodies to provide protection against all the hundreds of types of common cold virus.

The most difficult challenge to our immune system is posed by the AIDS (Acquired Immunodeficiency Syndrome) virus. This virus attacks the helper T cells and disables the immune system of its victims at the outset. Nevertheless, many people have succeeded in producing antibodies against the virus without as yet developing the disease. It may be that many of the victims of AIDS have already had their immune systems weakened by other diseases, which may help to explain the rapid spread of the virus in some parts of Africa and its slower spread in other parts of the world. The course of the spread of this disease is still shrouded in uncertainty, but the fact that it has not spread with the rapidity of the great bacterial plagues of the past suggests that the body can in many cases defend against it.

Given the chance, the immune system can meet all these challenges and many more. It has evolved to its current state of complexity and sophistication because there has been strong selection acting on it every generation. People living in Western societies over the last two or three generations have probably been the only ones in all of human history who have not had to use every bit of the immune system's capacity simply to survive.

THE STRUCTURE OF THE IMMUNE SYSTEM AIDS ITS EVOLUTION

The joining together of V, J, and C genes results in a change in the genetic information of the cell. If we define a mutation as just such a sudden change that can be passed on to the cell's descendants, then these alterations certainly fall into the category of mutations. The survival of the organism is also affected, even though the changes are passed on only to other somatic cells and not to the next generation.

But unlike the mutations caused by extrinsic agents that we talked about in the last chapter, these mutations are remarkably specific. During the development of each antibody-producing cell, similar sets of changes take place; these involve removal of similar-sized pieces of DNA. This occurs because there are enzymes, as yet poorly understood, that recognize specific sequences of DNA and remove them from the chromosome. They do so in such a way that it is not possible to predict which V, J, or C sequence will be inserted into the final gene, or exactly where the joins will be made. In this sense they are like random mutations. But they are not entirely random, because the entire structure of these long sequences of DNA and its attendant enzymes has been arranged in such a way that this process of building a gene and then modifying it through subsequent mutations is greatly *facilitated*.

We will soon be able to determine the precise sequence of events that has brought about the evolution of this elaborate system and of the mechanisms that continue to modify it. There is already good reason to suspect that viruslike or IS-like pieces of DNA may have been involved in its initial evolutionary stages.

Such bits of DNA are also probably involved in the changes that occur in the immune system at the present time. You will recall that the ''fine-tuning'' of the antibody is brought about through somatic mutations in the V region. These mutations are totally different from the random mutations caused by chemicals or radiation. Instead of occurring at random times and places, they happen in very specific parts of the V gene segment and only at certain times during the maturation of the antibody-producing cell.

Some sort of mutagenic agent or agents must produce these mutations, though we as yet have no hint of what they are. They cannot act unless the cells are entering the particular developmental sequence that leads to a mature antibody-producing cell. They also cannot act in the germ line, or in other cells of the body, even though they are present there, but must lie in wait until a particular signal is received. They are mutagenic agents because there is a large element of uncertainty about just exactly what kinds of changes they will produce in the V region. But, aside from that uncertainty, their mode of action is remarkably nonrandom.

Because the action of these agents is so precise, they begin to sound like some of the viruses and IS sequences we talked about in the last chapter, which can insert themselves into specific parts of the genome. We will see later that small mutations, such as those produced in the V genes, can also be made by transposable elements in the course of their entering and leaving the genome.

We are already beginning to discover other mutational agents with similar properties. In chapter 9 we will look at some remarkable transposable elements in the fruit fly Drosophila. In this fly, the elements can only move about and make mutations in the cells that are passed on to the next generation. This is quite the mirror image of the antibody situation, in which the events are confined to certain somatic cells.

Antibodies vividly illustrate how evolution can be facilitated through the interplay of genomic organization and co-evolved mutational agents. The forces shaping this system have been those of individual selection. Now, let us turn to another system in which the genes specifying particular molecules have become organized for rapid evolutionary response. This organization, unlike the organization of antibody genes, took place in the remote past, but even so it is still possible to see how it was shaped by selection on those ancient individuals.

6

An Evolutionary Toolbox

Is hemoglobin a kind of oxygen tank? No, it turns out that it
is a kind of molecular lung. The hemoglobin molecule is an
organ in miniature.
—MAX PERUTZ in HORACE FREELAND JUDSON,
The Eighth Day of Creation

The Harvard evolutionist Stephen Jay Gould has often remarked that he was led into his field by a boyhood fascination
with dinosaurs. Many evolutionists have followed the same
developmental route. Even as adults we continue to be fascinated by that distant past. Were dinosaurs really warm-
blooded? Were they a uniform green or brown or did they
come in a riot of colors? How quickly did they move? How
complex was their behavior? Were they silent like most present-day reptiles, or did they hiss and scream like creatures
in a monster movie?

The last hisses and screams (if there were any) faded in a
blast of scalding air from a meteorite impact sixty-five million
years ago. We are left with tiny fragments of that time and
must use all our ingenuity and imagination to try to reconstruct it.

The reconstruction becomes more accurate with each passing year. We cannot yet answer all the questions I have just
posed, but it may soon be possible. Colors are a problem,
but sophisticated biochemistry may be able to pick up traces

of pigments in the fossil impressions left by dinosaur skin, of which there are quite a few. So far as dinosaur songs are concerned, it is possible by looking at skeletal remains to infer that many dinosaurs had respectable vocal equipment. Some of the duck-billed dinosaurs had complex air passages in crests on their heads that may have aided in vocalization.

Fossils used to be hauled out of the ground in the same way that Heinrich Schliemann smashed through the fragile layers of Troy. But we have since discovered that much more can be learned from painstaking examination of the fossil sites. As a result, workers have found intriguing suggestions that the duck-billed dinosaur *Hadrosaurus* took care of its young in the nest. Nests with young of a variety of ages have been found, which would not be the case if the young scuttled away immediately after hatching, as turtles do today.

If the dinosaurs were not warm-blooded, they certainly had a remarkably high metabolism. The spacing of tyrannosaur tracks indicates that these giant carnivores could run at up to fifty kilometers per hour, twice as fast as a human. The structure of their bones suggests a plentiful blood supply, like that of mammals. Deep sockets in the pelvis to accommodate the leg bones show that even the largest dinosaurs were capable of quick movement. Robert Bakker has demonstrated that the population of herbivorous dinosaurs supported relatively few carnivorous ones, as would be expected if each carnivore had to eat a lot of herbivores to stoke hot metabolic furnaces. Among cold-blooded animals, in contrast, there tend to be fewer herbivores per carnivore.

No matter how much we learn, we will probably still be wide of reality. But in science, if a question can be posed, no matter how far-fetched, it can eventually be answered. New techniques make the world of the dinosaurs much more approachable. The questions posed by youths who fall under the spell of worlds long vanished will be answered in ever-greater detail when those youngsters grow up and, as the new generation of evolutionists, come to grips with the reality behind their childhood dreams.

One of the most exciting aspects of science is that it has provided us with time machines in which we can return to the distant past. Fossils are one such time machine. The images of distant galaxies, sending us light from billions of years ago, are another. A third, with which the reader is probably much less familiar, involves the use of giant molecules.

Because all the organisms on the planet are related, their giant molecules are related as well. And the molecules give us far more information about their family trees than do their current appearances.

Consider horses and yeasts: they certainly had a common ancestor, but the most cursory inspection will show that there are clear-cut differences between them. Yeasts are microscopic, single-celled organisms that reproduce by budding. Horses are macroscopic and multicellular, and reproduce in a considerably more complex fashion. It is thus not possible by examining these two species, or indeed by examining all the organisms alive at the present time as well as all of the known fossil record, to infer what the properties and appearance of the common ancestor might have been. Too many of the intervening creatures have disappeared and left no living unchanged descendants or traces in the fossil record.

All organisms, however, share very similar DNA and proteins, which consist of linear arrangements of subunits. Thus, it is possible to infer a great deal about the probable sequence of the molecules possessed by their common ancestors. We may not know what the organisms looked like, but we can infer what some of their molecules were like.

The horse and yeast alcohol dehydrogenases are an excellent case in point. The yeast alcohol dehydrogenase is 347 amino acids long, as I mentioned in the last chapter, and the horse alcohol dehydrogenase is 374 amino acids long. Figure 6.1 shows some data, kindly provided by Hans Jörnvall of the Karolinska Institute in Stockholm, that show how it is

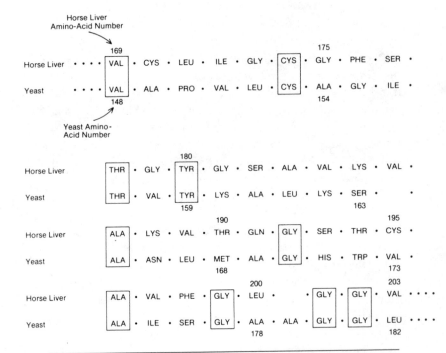

Figure 6.1 A comparison of part of the sequences of horse liver and yeast alcohol dehydrogenases, extending from amino acids 169 to 203 of the horse liver sequence and 148 to 182 of the yeast sequence. Enough of the sequences are shown to give you an idea of how such comparisons are made. You can see that only about a quarter of the amino acids (marked by boxes) are the same, but that this similarity is enough to allow the sequences to be lined up even though small deletions and insertions have been made in the course of evolution. The best guesses for two of these are shown: a deletion of one amino acid in the yeast sequence following position 164, and another single amino acid deletion in the horse liver sequence following position 200. Of course, these mismatches might have been caused by insertions in the other sequence instead. The similarities between the two sequences are not great, but are remarkable when you remember that these molecules have pursued different evolutionary paths for billions of years. (Modified from figure 1 of H. Jörnvall, B. Persson, and J. Jeffery, Characteristics of alcohol/polyol dehydrogenases, *Eur. J. Biochem.* 167 [1987]:195–201.)

possible to match up the two sequences. To see the resemblance between the sequences, it is necessary to suppose that short deletions and insertions of a few amino acids occurred during the two to three billion years during which they evolved separately. When these slight modifications are made, it is possible to match the sequences up in such a way that about a quarter of the amino acids are the same. The common ancestor of these two molecules was different from either of them, but it did share many amino acids with its remote descendants. Although a 25 percent shared identity may not seem to be much of a resemblance, it is remarkable that there should be any trace of a resemblance at all after such an enormous span of time.

Further, the resemblance is greater than it first appears. Since some amino acids are much bulkier than others, the long chain of amino acids in figure 6.1 is actually more like a necklace of different-sized beads. In its native state the "necklace" is coiled and folded back on itself in a highly specific fashion, providing the protein molecule with a three-dimensional structure.

Forming this three-dimensional structure as the protein is synthesized is equivalent to fitting such a necklace into a small jewel box with a highly convoluted and specific shape. If in the course of evolution a small bead is replaced by a larger one, the necklace might no longer fit unless a large bead nearby is changed to a smaller one. But because of the convolutions of the necklace, the second bead that must be changed to preserve the fit might lie a long way away from the first along the linear chain, even though it is right next to the first in the three-dimensional configuration.

Jörnvall showed that exactly this kind of pairwise accommodation has occurred again and again in the course of the evolution of these two molecules. This indicates that the three-dimensional shape of the two molecules has been preserved to a much greater degree than the 25 percent amino acid identity would indicate.

This pattern of conservation of three-dimensional structure in the course of evolution has been found for many other protein molecules. Protein structures are determined by a technique called X-ray crystallography, one of the most difficult and technically demanding endeavors in all of science. For a protein structure to be determined by this technique, it is necessary first to grow a large crystal of the protein. Tiny microscopic crystals will not do. The crystal must be at least a millimeter or two on a side, so that it can be handled and the X-ray beam directed through it.

If a satisfactory crystal is grown, the crystallographer must then carefully dope it with a scattering of heavy atoms to serve as reference points. Finally it is ready for the beam of X-rays, which is shone through it in different directions along its various planes. The X-rays are scattered from the ordered ranks of atoms in the crystal. Complex calculations using the patterns of these reflections can give, with great precision, the positions of the thousands of individual atoms in the molecule.

This technique requires large amounts of available protein, which until recently meant that only abundant proteins could have their three-dimensional pictures taken. Now, of course, it is possible to transfer any gene to bacteria, and soon it will be possible to make any protein in abundance, no matter how rare it is in the original organism. But even these cloned proteins still have to be crystallized, and only a small proportion of the proteins that have been studied make satisfactory crystals.

In spite of all these difficulties, a great many three-dimensional protein structures have been determined. The first group to have been successfully crystallized was the hemoglobin family, all the members of which turn out to be clearly related. Insect hemoglobin, for example, closely resembles mammalian hemoglobin in its three-dimensional structure, even though only 14 percent of their amino acids are the same.

This maintenance of an overall resemblance when most of the individual parts have changed is rather like the evolution of the watch. The first watches were made of brass, steel, and enamel and were sometimes encased in precious metals. The movements of the hands were controlled by a complex collection of wheels and gears, driven by an escapement that released the power of a coiled spring in small uniform packets. Polished jewels were sometimes used as bearings to reduce wear on the mechanical parts. Most present-day watches, like the one I am wearing, are primarily made of different materials—silicon chips, plastics, and organometallic compounds. They contain glass, and some of the more expensive ones have stainless steel backs, but that is the extent of their resemblance to old-fashioned watches. They have no moving parts, and if they appear to have hands this is because their movement is cleverly simulated by means of a liquid crystal display. Yet they are recognizably watches, and they have a function similar to the mechanical originals.

Not all the genes of yeasts and horses can be compared as directly as their alcohol dehydrogenases. Horses have many enzymes and other proteins that yeasts do not—hemoglobin is one obvious example. But all these extra genes must have come from somewhere, and one of the most exciting things about the study of molecular evolution is that we can begin to glimpse their origins.

THE GENEALOGY OF THE GENES

Enzymes need to be large and complicated in order to do their job of facilitating reactions between smaller molecules. When an enzyme has been sufficiently honed by evolution, it can make a reaction proceed at blinding speed. Carbonic anhydrase is a remarkable example. As the blood passes through respiring tissues where CO_2 gas is being generated, this enzyme takes the CO_2 and water and puts them together to form the soluble molecule carbonic acid. The blood can then carry the solubilized CO_2 away from the tissues. In the capillary beds of the lungs the process is reversed and CO_2 gas is released. One molecule of carbonic anhydrase can produce over a half million molecules of carbonic acid per second, which makes it the fastest enzyme known. It was the runaway activity of this enzyme in the drastically reduced blood volume of his wounded patients that Walter Cannon was fighting when he injected them with buffer.

Enzymes can build molecules up or break them down. They are responsible for making all the proteins of the body, even including more copies of the same enzymes. And after death they can escape from their normal functions and aid in the destruction of the body. If you let a piece of venison hang for a few days, it will become much tenderer. This is because enzymes are breaking down the connective tissue that gives the fresh meat a stringy texture. Of course, at the same time other enzyme-catalyzed reactions are taking place, breaking muscle proteins down into molecules that only a true venison aficionado could love.

The family trees of enzymes can be extremely complicated. The alcohol dehydrogenases found in mammals and yeast are quite similar to each other, as we have seen. Yet they have no detectable resemblance to two other enzymes with the same function, those found in Drosophila and in certain bacteria.* The Drosophila and bacterial enzymes also have no detectable relationship to each other. It really looks as if the genes for these various enzymes have had different evolutionary histories for a very long time but eventually converged to perform the same task in these very different organisms.

---

* Parts of these enzymes do have a family relationship to each other, but this relationship is so remote as to be at the very limit of detectability. The amino acids show no resemblance, but there are hints that the three-dimensional structures of parts of the molecules are somewhat alike.

Drosophila is a much closer relative of ours than yeast. Why should fruit flies and ourselves have different enzymes for this task? Did the gene for the Drosophila enzyme perhaps come in from another organism, somehow shoulder aside the original gene, and take over its job? This is not as farfetched as it seems, for it has recently been found that yeast has an extra gene for alcohol dehydrogenase that closely resembles the one found in the bacteria.

Thousands of different genes have a similarly convoluted history. Their family trees promise to be more complex than anything you will find in Debrett's or the Almanach da Gotha. Yet, as we gather more and more of these genealogical threads into our hands, it should be possible to see an emerging pattern. The goal is to push the story of protein evolution as far back into the past as possible, to get a glimpse of the time when there were relatively few enzymes in a few very simple organisms, that is, when every enzyme had many functions—like the butler Max (Erich von Stroheim) who did everything for the aging Gloria Swanson in the movie *Sunset Boulevard*.

How many enzymes were there in the primitive cell? Perhaps only a few dozen. The simplest organisms we know of that are capable of growing and reproducing independently are tiny bacteria-like creatures called mycoplasmas. (But they are only quasi-independent, since they must live in very close association with animal or plant cells in order to survive.) The smallest mycoplasmas have enough genes to make about 600 or 700 proteins. Many of these are building blocks for the cell itself, but at least half are enzymes. If we assume that many of these enzymes are responsible for the present-day characteristics of the mycoplasmas, and not therefore strictly necessary, then we can perhaps pare the number of absolutely essential enzymes down to one hundred or even fewer.

This small number becomes more likely if we remember that at the time life evolved, the first primitive cells were the only organisms around. They had the whole world to themselves. Not only did they perhaps require rather few enzymes in order to grow and reproduce, but those few did not need to be terribly efficient or even highly specific.

Now, if this is all true, then we would expect to see something quite remarkable. We would expect to see that all present-day enzymes, all the many thousands of them, are variations on a few basic themes. They should all have their ancestors in that small, irreducible number of enzymes doing their clumsy best in a simple cell just a few steps removed from the primitive soup.* By looking at the relationships between diverse families of enzymes, it should be possible to see back in time more than three billion years to somewhere near the very beginning of life.

In fact, what we see is something even more remarkable. It is certainly true that we can begin to group all the diverse enzymes of the present day into a relatively few families. But we also find that enzymes within a family are often built up of an even smaller number of building blocks, arranged in different ways in each enzyme. This gives a glimpse back to a time when there may have been even fewer than one hundred genes. And the ways in which these building blocks have been put together in the course of evolution show a striking resemblance to the ways in which antibodies are put together in our own cells.

Many families of enzymes have evolved like a set of snap-on tools. In using such a set, one starts with a handle and snaps on a variety of gadgets to end up with a hammer, a wrench, a screwdriver, and so on. When we find a system like this in the natural world, we cannot doubt that such an

---

* The primitive soup was a mixture of organic molecules, produced by chemical processes rather than living organisms. These molecules were assumed to have accumulated in the early oceans or perhaps in small ponds, from which living organisms are presumed to have arisen. It now appears that this soup may not have existed, or at the least was much more watery than we imagined a few years ago. As a result, theories of the origin of life based on hybrids of organic and inorganic catalysts are now beginning to gain favor.

arrangement has facilitated the evolution of the great diversity of present-day enzymes.

GLYCOLYTIC ENZYMES AS SNAP-ON TOOLS

Biochemistry as a science began in 1893, when Eduard Buchner and Martin Hahn carried cakes of brewer's yeast to their primitive laboratory in a basement at the University of Kiel. Their idea was to try to extract the cell sap from them and examine its properties, but they had no clear conception of what they would find.

Buchner, who had already worked with yeast, had discovered that Louis Pasteur was wrong when he stated that yeast would only ferment in the absence of oxygen. Buchner had found instead that the cells would continue to ferment even when oxygen was present, so long as there was plenty of sugar in the fermenting juice.

The yeast cell, simple as it was, was a black box. Pasteur felt that there was something about its being alive that permitted fermentation to take place, for he found that dead cells completely lost their ability to ferment. The question of whether there really was such a vital principle had torn the fledgling field of biochemistry apart.

When Buchner and Hahn found that they could grind yeast with a mixture of sand and diatomaceous earth, then wrap the resulting mush in cloth and squeeze it in a press to expel a clear yellow fluid, they had no idea that they were about to answer that question. Their first experiments were failures; bacterial action quickly turned the fluid cloudy and flocculent. Chilling it on ice slowed the process down, but did not stop it.

Then Buchner hit on the idea of trying to preserve it the way jams and jellies are preserved, by adding a strong solution of cane sugar. When he did this, the fluid began to foam and bubble furiously. He found to his astonishment that it was making carbon dioxide and alcohol just as the living cells did.

That moment spelled the death of Pasteur's vital principle. It also opened up a world of biochemical reactions that could be carried out in the test tube.

Buchner came from a poor family, and though he continued to be productive, his subsequent career forms a sad commentary on the way the German Empire treated its distinguished subjects. When he won the Nobel Prize in 1907 for his discovery of cell-free fermentation, he was finally given a proper teaching post at a decent salary. But this part of his career was brief, for he volunteered when war broke out in 1914. The army used him for a variety of dangerous jobs, and he was mortally wounded by shrapnel on the Rumanian front in 1917. The German army's casual wastage of this Nobel Laureate gives new meaning to the term *cannon fodder*.

In the meantime Buchner's work had sparked an explosion of excitement among scientists. He thought at first that his material, which he called *zymase*, was made up of a single component. Others quickly discovered that it was really a diverse family of molecules, eventually called the *glycolytic* (sugar-dissociating) enzymes.

These enzymes break down sugars, on which we depend for much of our energy, into simpler molecules. The energy released is used to make new high-energy molecules called NADH and ATP. There are about a dozen glycolytic enzymes, each responsible for a single step in the process.

These enzymes are very old, because the process of glycolysis is very old. We can trace them back to a time long before the hemoglobins, indeed before there was any free oxygen in the atmosphere for the hemoglobins to pick up and carry. The simplest primitive organisms must have carried out glycolysis, using enzymes very much like those we have today. We can be fairly confident of this because glycolytic enzymes have been highly conserved during evolution: the glycolytic enzymes of bacteria are 50 percent identical with our own.

We might expect that a diverse group of enzymes that performs a variety of tasks would have many different structures. This turns out to be only partly true. To see this, we must understand something about the structure of such globular proteins, which I described a moment ago as folding back on themselves like strings of beads in a jewel box. If you tried to construct a protein by stringing together amino acids picked at random, you would find that it would collapse on itself in a fairly meaningless jumble. But proteins that have been shaped by the process of natural selection tend to be made up of two types of structure.

The first is a winding helix, called the alpha-helix, which was discovered by the remarkable chemist Linus Pauling. (This is not the same as the famous double helix of DNA.) Sometimes quite long stretches of an amino-acid chain may form an alpha-helix, giving rise to tubelike pieces of protein. The tube may bend or have kinks in it because the helix is usually not perfect.

The second type of structure is a zigzag. Side groups of amino acids stick out on alternate sides of the zigzag and tend to lock together with those of nearby zigzag chains to form a sheet of amino acids called a beta pleated sheet. The chain of amino acids may double back on itself many times to form quite a large structure (see figure 6.2).

Amino acids have acidic and basic parts, and in a protein the basic part of one links to the acidic part of the next. As a consequence, the entire protein molecule has a polarity from its basic to its acidic end. This gives the chain of amino acids in each part of the sheet a specific orientation, shown as an arrow in figure 6.2. You can see in the figure that in the sheet found in each half of the molecule, all the chains are pointing the same way. This does not need to be the case. Depending on how the backbone of the molecule wanders about, the

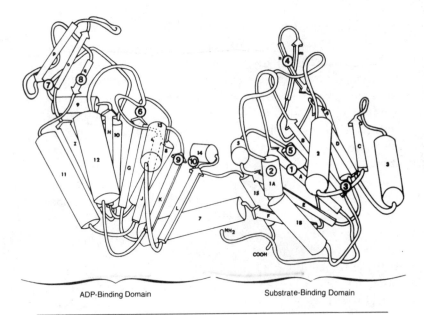

ADP-Binding Domain   Substrate-Binding Domain

Figure 6.2 Some fascinating features of molecular evolution can be seen in this diagram of the three-dimensional structure of a glycolytic enzyme, phosphoglycerate kinase, isolated from horse muscle. What appears at first to be a monstrous tangle resolves itself into interesting patterns on closer examination. You can see that the enzyme is divided into two domains. The one on the left binds the compound ADP, which is converted to a much higher energy compound ATP in the course of the reaction. The domain on the right binds the substrate for the enzyme, which donates a phosphate group to the ADP. This is one of the key energy-generating steps in the breakdown of sugars.

Now, look closely at each domain. The regions of alpha-helix are represented by cylinders, and the regions of beta pleated sheet are represented by flat arrows. You will see that in each domain there is a fanlike set of six arrows, all pointing the same way. This structure is typical of domains that bind high-energy compounds. These and other similarities between the domains—note the two arrows pointing in opposite directions in the top part of each domain—suggest that the two domains of this enzyme resulted from a gene duplication in the distant past. There are some structural differences between the domains, and the introns (represented by circled numbers) are at different places in the two domains. Nonetheless, an excellent case can be made for gene duplication followed by subsequent divergence of gene function. (From figure 3 of A. M. Michelsen, C. C. F. Blake, S. T. Evans, and S. H. Orkin, Structure of the human phosphoglycerate kinase gene and the intron-mediated evolution and dispersal of the nucleotide binding domain, *Proc. Natl. Acad. Sci. U.S.* 82 [1985]:6965–69.)

chains can have either orientation with respect to each other and the edges of the zigzags will still lock together. It must be emphasized that a structure as complex as this particular sheet, with its six doubled-back chains and beautifully interlocking edges, is not likely to have arisen more than once. If we see pleated sheets with this specific orientation in two different proteins, or as in this case in the two halves of a protein, we can infer that they almost certainly had a common ancestor.

All globular proteins are made up of combinations of alpha-helix, beta pleated sheet, and short stretches of amino acids that do not fit into either category. Combinations of these structures form *domains,* substructures of the larger protein molecule. Just as we can look for amino-acid identities between remotely related molecules, we can also look for domain identities.

Relationships between superficially very different enzymes can be detected if their three-dimensional structures are compared. Michael Rossmann and his co-workers at Purdue University have compared the patterns of folding in the glycolytic enzymes. They find that this diverse collection of enzymes is made up of a small number of domains arranged in different ways in the different molecules. This can be seen in figure 6.3.

Domains of the same type, while different in detail, are recognizable by the varying amounts and arrangement of their components, alpha-helix and beta pleated sheet. The most common building block that is found among the glycolytic enzymes is that six-stranded beta pleated sheet, depicted in Figure 6.2, in which all the strands have the same orientation. It is possible to imagine some great advantage to a six-stranded sheet, so that it might have evolved more than once. But it stretches the bounds of credibility to expect that all six of the strands in such independently evolving structures would always be pointing the same way.

This particular building block has different functions in different enzymes, but all involve energy transfer. One of its most important functions is to bind a molecule called NAD, which can then pick up energy from the small piece of glucose that the enzyme is working on. After this the NAD is released in an energy-rich form called NADH. The six-stranded building block is a part of so many enzymes because such energy transfers are an excellent way to perform many important biological reactions. This block is analogous to the handle of a set of snap-on tools.

ROUGH CARPENTRY OF THE GENOME

Remember that the first enzyme genes were few in number and probably coded for molecules capable of catalyzing a broad range of chemically similar reactions. This would not result in particularly efficient organisms, since such an enzyme is unlikely to be particularly quick or accurate in its action. As organisms became more specialized, there would be strong selection for increased numbers of genes. This in turn would allow each gene to become more specialized.

As with the antibody-coding genes, this might have happened in one of two ways. One possibility is that new genes

---

Figure 6.3 (*overleaf*) This diagram shows how a variety of enzymes have been built up from a relatively small number of building blocks. The triangles represent regions of beta pleated sheet, and the loops represent lengths of alpha-helix. You can see that in the course of the evolution of these enzymes the building blocks have been shifted about, duplicated, deleted, linked together in a variety of ways, and sometimes reversed. Most of the time, the beta pleated sheets have retained a particular orientation, as in figure 6.2. In the diagram, this orientation is represented by an upright triangle. (From figure 13 of M. G. Rossmann, A. Liljas, C.-I. Bränden, and L. J. Banaszak, Evolution and structural relationships among dehydrogenases, in *The enzymes*, ed. P. D. Boyer [New York: Academic Press, 1975], pp. 61–102.)

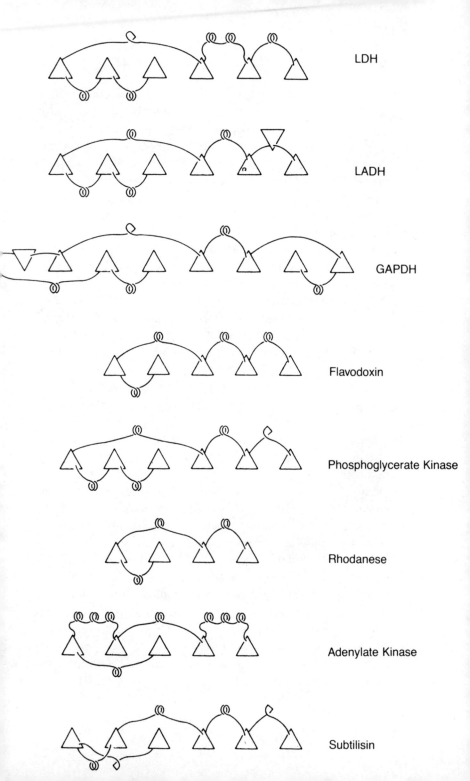

LDH

LADH

GAPDH

Flavodoxin

Phosphoglycerate Kinase

Rhodanese

Adenylate Kinase

Subtilisin

could arise by duplication and slowly evolve to take up new functions. But this tends to be a very lengthy process, and it would be very difficult for a new enzymatic function to arise quickly in response to a very pressing need for it. A second possibility is that genes or parts of genes that had already been selected for particular functions could be quickly brought together to perform new functions, either as a result of unequal crossing-over or through the agency of jumping genes.

Suppose there were a great advantage to performing a new kind of reaction but that no organism alive at that time made an enzyme that could carry it out. If the energy of the reaction could be transferred to NAD, then the first step in evolving such an enzyme might be the chance shifting of a copy of the NAD-binding gene to some other stretch of DNA elsewhere in the genome. This DNA segment might not have had a function up to then, or it might have coded for a protein with quite a different function. After the insertion of the NAD-binding gene, however, the protein segment would suddenly be part of a new molecule with NAD-binding properties.

While new proteins made in this way usually had no detectable new function, every once in a while one would arise that could carry out the new reaction. Clumsily perhaps, but many times more quickly than the reaction would occur spontaneously. This is not as surprising as it sounds. Random pieces of protein constructed in the test tube have been shown to catalyze a variety of different chemical reactions, though many orders of magnitude less quickly than a properly evolved enzyme. Further, it must be remembered that this new gene was not simply a random piece of DNA. Half of it was, true, but the other half was a highly evolved gene that made a protein capable of binding NAD and positioning it where it might aid a new reaction.

Once the new enzyme was born, it was quickly improved through mutation and natural selection. The organism carrying the new gene had a great advantage over its fellows.

The use of the NAD-binding piece as a building block for the new enzyme had immediately solved half the organism's problem. From the very beginning the new gene worked far better than if it had to be built from scratch by natural selection.

It may be that we can still see traces of the points at which these bits of gene were joined together. When the genes of higher organisms first began to be sequenced a little more than a decade ago, it was quickly discovered that they were not continuous, uninterrupted stretches of message-bearing DNA like most of those in simpler organisms. Instead they were interrupted, sometimes in many places, by stretches of DNA that do not carry information for making proteins.

You will recall that not all DNA codes for the production of proteins and that there are long stretches of DNA that are not genes in the usual sense. Some of these long stretches are the ones that interrupt the gene and make it much longer than it needs to be, as much as a hundred times longer.

These long stretches are called introns, and the bits of coding sequence that they interrupt are called exons. All these long stretches of DNA, along with the exons that they separate, are turned into RNA. The introns are then removed by specific enzymes to produce the mature message that is translated into protein.

Why this complicated and apparently quite unnecessary exercise? It is like an article in a Sunday newspaper that the patient reader pursues from page 1 to page 37 to page 62 to page 11E in section D8. Delete all the intervening ads, and the whole thing would take up a couple of columns.

A possible reason for all these introns was quickly suggested by Walter Gilbert, one of the inventors of ways to sequence DNA rapidly. Many of the introns appear to be positioned at or near the boundaries between different domains in the protein molecule. Might they aid in the process of building new genes by somehow facilitating swaps of one exon for another?

This presumptive process has been christened _exon shuffling, and it has generated a good deal of excite_ment. There are some problems with it, however. First of all, primitive and simple organisms have few introns. This may be because introns appeared relatively late in evolution. Or it may be that the earliest organisms had them but that they were lost in simple creatures like bacteria because such creatures needed streamlined genomes. If introns are of recent origin, then it must have been possible for exon shuffling—or more properly domain shuffling—to take place before the appearance of introns. If they are old, this leaves unanswered the question of why we higher organisms have retained our introns. Carrying this enormous extra baggage of unnecessary DNA each generation seems a heavy price to pay for the privilege of the occasional bout of exon shuffling.

Second, it is unclear what the mechanism of exon shuffling might be. Some introns resemble transposable elements, but most do not. Introns certainly change with blinding speed compared with most other evolutionary processes, and we have not yet discovered how they do it or whether they can move exons around in the process. Third, there are so many introns in most genes of higher organisms that it is not surprising many of them fall on or near the boundaries between functional domains. It may simply be a coincidence.

It would be nice to be able to decide this question by looking for very old introns in the same way that we look for very old genes. Unfortunately, introns change so quickly that the introns of even closely related organisms often show no resemblance to each other beyond the fact that they occupy the same positions in the gene. The jury is still out on exon shuffling per se, but it is well to remember that even if introns play no role in such gene scrambling (though they almost certainly play at least some role) _domain_ shuffling has still taken place. We have firm evidence that this has happened from the sequences of the genes involved.

It is, I think, far from coincidental that the process of domain shuffling begins to sound very much like the antibody story from the last chapter. But rather than happening each generation, the genetic events that shaped the glycolytic enzymes were spread over millions of generations.

We can see from these examples of domain shuffling how both individual and species selection must have worked together in the evolution of the glycolytic enzyme toolbox. While there was a great advantage to the individual in the gaining of a new biochemical step, there was also an advantage to the species that had the best toolbox. *Best* in this instance means two things: the toolbox must have been well organized and accessible, and it must have contained not only the pieces of gene that could be cobbled together but also the mutational agents that were most likely to cobble them together in useful ways.

The evolution of building-block domains made the appearance of new enzymes more likely, just as the organization of antibody genes makes the appearance of new and useful antibodies more likely. Protein domains form an evolutionary toolbox of extraordinary utility.

The glycolytic enzyme toolbox is probably closed and covered with dust now. Since the glycolytic pathway is firmly embedded in the genes of all organisms, anything that changes it dramatically is almost certain to be harmful. The mutagenic agents that put its domains together in new ways must have been long banished from that part of the genome. But when we look at these enzymes, we can still see traces of a time when the world was young and modification of the genome through a kind of rough carpentry was enough to give an organism an advantage. New and much more sophisticated toolboxes have been opened in the meantime.

But you are probably reeling by now from all these molecular complexities. Let us take a bit of a breather and see how evolutionary facilitation might work at the level of organisms rather than that of molecules. Let us, in short, enter for a little while the airy world of butterflies.

7

Of Butterflies and Handbags

The model Monarch as a caterpillar eats
Milkweeds, stores their hearty poisons,
   and the butterfly defeats
Some portion of an avian predation team
By advertising—with show of color, lack of haste—
That those who dare to peck will find it in bad taste.
              —JOHN M. BURNS, *Up the Food Chain*

No group of organisms in the living world is as subject to the fierce and varied pressures of evolution as the insects. As a result, they have had phenomenal success. The insects were some of the first colonizers of the dry land, and they now make up over three-quarters of the described species of animals.

Only one small group of midges has learned to reenter the ocean, but elsewhere insects are omnipresent. The larvae of other tiny midges grow in ephemeral pools of water during the brief Antarctic summer. Flies multiply happily in the sumps into which gas station attendants drain used oil from our crankcases. Bugs skitter in surprising number and variety across the sea's surface.

The extraordinary success of the insects is in part a function of the arrangement of their genomes and the ways in which these genomes interact with the mutagenic agents that alter them. We will see some specific examples of this in the next chapter, in the discussion of recent work on insect DNA. The phenomenon of evolutionary facilitation is very plainly written on their genes, and getting plainer each year as information continues to pour in.

In this chapter, however, I would like to get away from molecules for a bit and ask how the process of evolutionary facilitation might look at the level of the organism. Are there some obvious examples of this facilitation in the world around us? In particular, since we now know something about the molecular mechanisms involved, can we find some examples among the insects?

There are in fact many. The evolution of mimicry in butterflies is particularly striking. In some cases, the ability of these insects to mimic others derives from the evolution of elegant and complex sets of genes that act as mimicry toolboxes. Although we know little as yet about the molecular nature of these genes, we do know enough to be able to speculate about how they may have evolved.

Mimicry must be a very old adaptation that probably first appeared as camouflage—cryptic coloration or behavior, or both. Disguising oneself to hide from predators is a sensible thing to do, and the ability to accomplish this must have evolved very early. The ability of organisms to mimic their surroundings has reached remarkable levels. Figure 7.1 shows some famous photographs of a flounder, a bottom-dwelling fish, striving with surprising success to mimic a variety of backgrounds, including checkered tablecloths.

But the evolution of this type of mimicry also stimulated the evolution of predators—in order to find the mimics, they were forced to evolve more and more sophisticated sense organs and the brains to use them effectively. This is a race that can only go so far. A determined predator will eventually track down the prey, no matter how still the prey crouches or how much like a leaf or a lump of fecal material it has evolved to resemble. Further, the necessity for disguise severely limits the prey's activities. An organism that spends

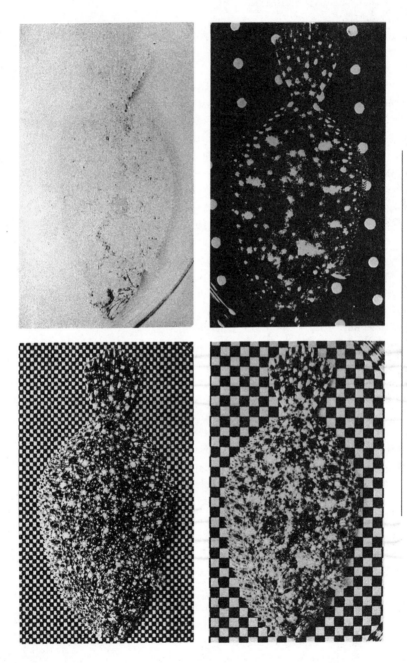

Figure 7.1 The flounder *Paralichthys albigattus* is able to take on an astonishing variety of patterns. Here you see the same fish against four different backgrounds. The process of adaptation takes a few hours in each case.

all its time hiding has a limited existence. The world would be a colorless and uninteresting place if it were nothing but a vast game of hide-and-seek.

Some mimics have been able to free up their own life-styles by imitating parts of the environment their predators might find objectionable, such as other species that are distasteful or dangerous. If the species being mimicked is brightly colored or numerous, as is often the case, this actually has the effect of turning the predator's sophisticated sense organs and intellectual equipment to the prey's advantage. The predator can see the prey easily but also learn to avoid it.

Other species have developed mimicry as a result of completely different selective pressures. For example, the flowers of a number of species of orchid mimic the female of their pollinating insect species. This increases the likelihood that they will be visited by males and pollinated. After a long period of coevolution they mimic not only the shape and color but also the scent of the female, and they may bloom before the real insect females have become numerous enough to compete with them for males.

Mimicry can often be extremely subtle. On coral reefs there are *cleaner stations* where large fish go to have their external parasites removed by smaller fish. Outside the cleaner stations, the small fish would be the natural prey of the larger ones, but in these safety zones they swim into their mouths and forage around their gills with impunity. The larger fish often line up at the cleaner stations, patiently waiting their turn like cars at a carwash.

The small cleaner fish are usually members of the wrasse family, and they are clearly marked with bright-colored horizontal stripes. But some small blennies mimic these cleaner wrasses in appearance and behavior, swarming around the cleaner station. Instead of eating parasites, the blennies take neat bites out of the fins of the fish waiting to be cleaned. They trade on the disinclination of the big fish to eat the wrasses that keep them clean.

There are very different limitations on the populations that practise different kinds of mimicry. Cryptic organisms that simply disguise themselves as part of their surroundings are limited in numbers by the amount of the habitat they imitate. If there is a good deal of this habitat, they can become common—though not so common, of course, that a predator can learn to flush them by simply stumbling around.

But if instead a mimic survives by imitating some model organism with striking properties, its numbers are dependent on those of the model. Indeed, it must usually be less common than its model, and in some cases much less. If the parasitic blennies multiplied to became an appreciable fraction of the cleaner fish population, the fish being cleaned would soon find the operation to be an unpleasant rather than a pleasurable one. They would quickly learn to munch on these tiny fish that had turned from helpful to irritating.

While mimicry is found in many other groups of animals and plants, it was studied in detail first in the insects. Many types of mimicry have been described.

Müllerian mimicry is named after Fritz Müller, a German naturalist of the last century who worked in Brazil. This very common type of mimicry, dependent for its success on the same reflex that we have developed in response to a stop sign, can be fairly simple or incredibly complex. In essence, it describes the fact that dangerous animals or plants will tend to converge on similar warning patterns or behavior. For example, many stinging insects such as bees and wasps have evolved a striking pattern of alternating black and yellow stripes, which immediately signals to a potential predator: "Don't eat me! I sting!"

This is indeed just like our familiar stop sign, which has become replicated, down to the word *Stop*, even in some non-English-speaking countries.* By now, the pattern of black

---

* This includes France, but not Quebec, where militant Francophones insist that it should say *Arret.*

and yellow stripes has become just as ubiquitous in the insect world as the stop sign has in ours. The Müllerian mimic is the beneficiary of a long history of convergence on a single pattern.

It does not much matter what the pattern is. Once it has been established as dangerous, many species may converge on it. Müller was originally struck by the similarities among several species of heliconid butterflies in South America, and suspected that this similarity was due to the distastefulness of the species being mimicked. Later workers have found Müllerian mimicry in butterflies to be widespread in the tropics, particularly in the New World, and to exhibit an incredible and exuberant complexity.

Groups of species can converge on a number of different patterns in the same area so long as patterns are clearly different from each other. In one part of central Peru, five strikingly distinct Müllerian patterns are found among the butterflies, leading the casual observer to assume that there are rather few species present in this forested area. But each group (called a *Müllerian ring*) can consist of dozens of species of several different families, all imitating each other and deriving mutual benefit.

There is nothing sacred about a given pattern. The set of rings in one area might be quite different in appearance from the set of rings in another, even though the same species may be taking part in both. *Heliconius erato* and *H. melpomene* are two closely related but clearly distinct species of heliconids. If these species are captured in a particular area, they are usually found to resemble each other closely and to belong to a particular Müllerian ring. But if members of the same two species are captured in another area, sometimes only a few miles away, they may belong to another and quite different ring and have quite a different appearance. The color and pattern differences can be so striking that the casual collector will assume they are different species.

The pattern that is being converged on is immaterial. Indeed it does not have to be particularly obvious (though being obvious probably helps). The important thing is that the predators in a given area have learned that these insects taste unpleasant. Experiments have been done to show that most of the members of a Müllerian ring really are distasteful to birds. But their degree of distastefulness varies. And some of the butterflies that have joined Müllerian rings are not distasteful at all.

These latter, nonpoisonous butterflies belong to another important class of mimics, named *Batesian mimics* after the naturalist Henry Walter Bates, who first described this type of mimicry clearly. Like Müller, Bates did most of his important work in tropical South America, traveling on some of his expeditions with our old friend Alfred Russel Wallace.

Not all Batesian mimics have joined rings. In many cases, a distasteful or dangerous organism will have its own unique color and pattern. If the organism is plentiful, it will sooner or later stimulate the evolution of Batesian imitators among less dangerous or more toothsome species. Examples of pure Batesian or pure Müllerian mimicry are probably rather rare. Within each model or mimic species there may be members with different degrees of distastefulness or dangerousness. But the categories are convenient, so long as we remember that they are not absolute.

Batesian mimicry of a single model is a precarious evolutionary route for a mimic to take for two reasons. First, there is safety in numbers. If a Batesian mimic slips into a Müllerian ring, it may be able to grow to quite substantial numbers without its disguise being penetrated. And while the distasteful members of a Müllerian ring may fluctuate in numbers, there will probably always be enough of one or more of these distasteful members around to ensure that the Batesian mimic continues to be protected.

But if there is only one model species, things become much less stable. The mimics are much more at the mercy of their environment and much more closely tied to the fate of the model species. They must resemble their model sufficiently closely that the predators are fooled. Ideally, they should behave in the same way that the models do, and of course they must be found in the same places. And while they can increase in numbers when the model is plentiful, if the model suddenly becomes rare they can be put in great jeopardy as they go on blindly imitating their absent protectors. A new generation of predators will quickly learn how delicious they are, and how easy to catch.

Even in the best of times, the mimics cannot become too numerous for two reasons. First, if they do their protection will be diluted. Predators will not meet enough of the models to learn to avoid them. Second, they may jeopardize the model as well. Too many toothsome mimics fluttering about means that too many models will get eaten—or at least munched on and spat out, which amounts to the same thing—by mistake.

We see exactly the same dynamics in human affairs. In Tijuana, Mexico, a brisk trade goes on in imitation Louis Vuitton handbags. (For readers who are not into the fashion scene, Louis Vuitton handbags are marked by a distinctive repeated pattern of interlocking monograms. They are therefore very easy to recognize and to imitate.) These imitation handbags, which flood in from the Orient, are also sold on the street in many American cities. As a result, sales of the original have plummeted. Who, after all, wants to carry around a handbag that will be taken by everyone for a cheap fake? It may be that in this case the mimics will drive the model to extinction.

The sensible response of the company making the overly mimicked model is to replace it with something quite different, and indeed new Louis Vuitton handbags do not resemble the old ones. The model butterfly can follow essentially the same route to escape from mimics that have become too numerous. Models exhibiting a distinctive pattern that appears as a result of a mutation might be at an advantage, since the predators will quickly learn to avoid them. If dra-

matic mutations occur at fairly high frequency in the population of models, some of them might survive the predators' initial learning period and spread through the population.*
This probably does not happen very often, but if it does, the mimic—left mimicking a model that no longer exists—will be in real jeopardy.

So the mimic must be poised to evolve at the drop of a hat, or at least at the appearance of a pronounced mutational change in the model. Even small changes in the model can keep the mimic on its evolutionary toes. Years ago, the British evolutionist and statistician R. A. Fisher pointed out that the model should always be altering its appearance a little, because this will hold the mimic population in check. The mimic must track the model as best it can. This has the effect of giving both mimic and model species a good deal of practice at evolving. We will shortly see what effect this has had on the organization of their genes.

Mimicry provides one of the most marvelously complex and at the same time accessible systems for studying evolution in action. Let us now look at some examples in more detail.

DINING ON HEARTY POISONS

One of the most remarkable and best-understood cases of Batesian mimicry in insects involves the monarch butterfly and its imitator the viceroy. The monarchs belong to the family Danaidae, members of which are usually both unpalatable and highly visible. They are named after the daughter of the king of Argos, Danaë, who was visited by Zeus in the form

---

* This is especially true if they are recessive, since they can spread for a while in the population before their effects on the phenotype begin to appear.

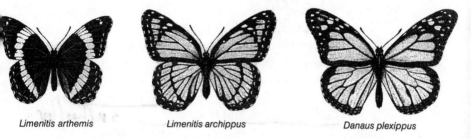

Limenitis arthemis   Limenitis archippus   Danaus plexippus

Figure 7.2   The most famous of all the mimic-model pairs. The butterfly on the right is the distasteful monarch, *Danaus plexippus*, and the one in the middle is its tasty and less peripatetic mimic, the viceroy, *Limenitis archippus*. The resemblance is even more remarkable when the colors as well as the patterns of these butterflies can be compared. The butterfly on the left is a close relative of the viceroy, with a markedly different wing pattern and much duller color. (From figure on p. 28 of L. P. Brower, Ecological chemistry, *Sci. Amer.* 220 (2 [1969]:22–29.) Copyright © 1969 by Scientific American, Inc. All rights reserved.

of a shower of gold and as a result conceived the hero Perseus. It is easy to see how they got their name: to be surrounded by a swarm of these butterflies is indeed like being in the middle of a shower of gold. Perhaps such butterflies were the source of the legend itself.

The viceroy belongs to quite a different family, the Nymphalidae (some of which are commonly known as fritillaries). These are named after their conspicuous larvae, which often have elaborate spines.

Even though they are only remote relatives, the monarch and the viceroy are remarkably similar in appearance, with a striking orange, white, and black wing pattern. This makes them visible a long way away, either at rest or on the wing (see figure 7.2).

The monarch is the most adventurous of insects. Various species are found worldwide, and each year new generations of monarchs undertake migratory journeys as long as those of many birds. As I write this, the last few of this year's migration of monarchs are fluttering by the window of my office on their way from the Great Basin states to breeding grounds along the California coast.*

Far more extensive migrations take place between the eastern United States, Canada, and central Mexico, where strenuous attempts are now being made to protect the breeding sites. On these journeys some of the butterflies can travel over 2,000 miles, at up to eighty miles a day. 3 n / l ↝/ un

Without a doubt, these insects could not survive these incredible journeys unless they had strong protection against predation. This protection takes the form of an extreme distastefulness. Monarchs contain high levels of calotropin and calactin, two compounds from a family known as cardiac glycosides. These extremely poisonous compounds alter the excitability of the heart muscle. (Note the pun on this fact in John Burns's poem at the beginning of this chapter.)

Birds that incautiously eat these insects are loath to repeat the experience. In some cases, they actually vomit them back up again. The taste must be particularly unpleasant, because once a bird has tried a monarch it usually will not try another. Even stupid birds learn after two or three attempts.

The glycosides, however, are not made by the monarch. They are made instead by the plants that the monarch larvae eat while they grow and develop during the northern summer. These are plants of the milkweed family, and their poisons protect them from many insects and from grazing animals. Since the monarchs can handle high concentrations of these compounds by storing them, they can benefit from an

---

* These butterflies breed in groves of trees to which their descendants will return year after year—until, of course, as is often the case, the trees are removed to make way for condominiums.

When I was a new arrival to California in the early 1960s, it was quite possible to find butterfly trees a few steps away from the coastal highway. The branches of the trees were laden with courting butterflies, and casual tourists could stop and shake a few thousand of them loose, breaking the romantic spell. Now, the few remaining trees are carefully fenced off and tourists kept at a safe distance.

adaptation that has evolved to counter quite a different type of predation.

One of the most conclusive experiments showing the protective nature of these compounds was carried out by the team of Lincoln and Jane van Zandt Brower, to whom we owe much of our knowledge about this fascinating mimicry system. Their beautiful experiment is one of the classic studies in biology.

With considerable difficulty, the Browers raised monarch larvae on cabbage rather than on milkweed. The adult monarchs were then fed to blue jays which had had no previous experience with the insects. The jays wolfed them down unhesitatingly, eating as many monarchs as they were given. In order to check that the hand-rearing of the jays had not somehow changed their behavior, the Browers then fed them monarchs caught in the wild. The jays found these distasteful and like their wild relatives quickly learned not to eat monarchs.

As an illustration that nature can be remarkably subtle, the Browers discovered that not all the monarchs caught in the wild were distasteful to the jays. In fact, about three-quarters of them were eaten without causing any discomfort. These particular insects had fed as larvae on nonpoisonous plants and were relying for protection against predation on their brothers and sisters who had ingested the cardiac glycosides. The Browers called this phenomenon *automimicry*.

Because of its bright colors, distastefulness, and active lifestyle, the monarch is an excellent model. As a result, both Müllerian and Batesian mimics have appeared. The best-known Batesian mimic is the viceroy, which belongs to an entirely different butterfly family and does not migrate. Because viceroy caterpillars feed on the leaves of various trees rather than on milkweed, the adults are perfectly palatable, as the Browers discovered by feeding them to the hand-reared birds. But when they tried to feed them to birds that had lived in the wild, the birds turned up their beaks in disdain.

Many members of the monarch family in different parts of the world show the typical orange-and-black pattern that serves as a warning of extreme distastefulness, but this may simply be because they are all related. Most of the closest relatives of the viceroy, however, have completely different wing patterns and much duller coloration than does the viceroy. Therefore, the pattern of the viceroy has certainly evolved as a result of a strong selective advantage for those of its ancestors that most closely resembled the monarch.

This is a dramatic evolutionary change on the part of the viceroy, and it leads us to ask how easily and quickly such a change could have occurred. It certainly has not happened in most of the viceroy's near relatives, although there are some viceroy relatives, particularly in Florida, that appear to have followed a similar evolutionary path to mimic relatives of the monarch.

Whether these different mimetic patterns evolved before, after, or during the divergence of these closely related species, we do not know. We also do not know how quickly this mimicry evolved, since we have no idea how long the mimic-model relationships have existed. But on the evolutionary time scale species have a rather brief existence. It seems certain that the viceroy and its relatives went through rapid evolutionary changes as they converged on the patterns of the monarch and its relatives.

Convergence towards a Müllerian pattern can also probably be quite rapid. The Müllerian rings of South and Central America probably evolved during the ice ages, when the tropical forest was more fragmented than it is now. And they may have evolved, or at least grown in numbers of member species, during the lifetime of those species that currently take part in them, since different groups within species like *Heliconius erato* and *H. melpomene* have evolved toward membership in a number of rings that differ widely in appearance.

But we can get a better feel for how delicately balanced the relationship between mimics and models can be, and how quickly it can change, by looking at the situation in sub-Saharan Africa. There are many beautiful examples of mimicry, but I would like to concentrate on two African mimetic species, *Hypolimnas dubius* and *Papilio dardanus*. These species tell us a very clear tale of evolutionary facilitation. I will deal with *Hypolimnas* first, since its story is more straightforward.

THE SWAMP DWELLER

Hypolimnas dubius is a nymphalid, like the viceroy. The name *Hypolimnas* suggests that this butterfly is found in marshy areas (actually it prefers mature forest), and *dubius* suggests that there is something doubtful about it. The name reflects an old taxonomic confusion: butterflies of this species come in two totally different forms, so strikingly different from each other that naturalists were fooled for a long time into thinking that *H. dubius* was actually two species.

H. dubius forms part of a complex web of Batesian and Müllerian mimicry that has taken years to untangle. The reason it has two forms is that in both eastern and western Africa it mimics two distinct model species. Two of the models that are found in West Africa, and the two forms of *H. dubius* that mimic them, are shown in figure 7.3.

Black-and-white reproduction cannot do justice to the differences between the two models, but it does give some idea of how closely the two mimetic forms resemble them. In addition to the very different pattern of white spots on the fore and hind wings, the right model in the figure has sulfur yellow rather than white hindwings, as does its mimic. The matches between models and mimics are quite uncanny. Both males and females of the two different forms of *H. dubius* are found in the same population, showing that their different patterns are not somehow tied to sex.

Amauris tartarea

Amauris niavius

Hypolimnas dubius

Two Forms of *Hypolimnas dubius* and Their Models

Figure 7.3 The bottom two butterflies in this picture are both *Hypolimnas dubius,* showing the two forms found in West Africa. The first form mimics two species of *Amauris,* and is almost a perfect match for *Amauris tartarea,* the butterfly you see above it. The second form mimics *Amauris niavius,* seen above it, as do some of the females of *Papilio dardanus,* found in the same area. Not only the appearance but the behavior of the models is mimicked, and *H. dubius* butterflies accomplish this with one or another allele of a supergene complex. (From figures on pp. 130–31 of D. F. Owen, *Camouflage and mimicry* [Chicago: Univ. of Chicago Press, 1982].)

In fact, the matches extend beyond appearance. The two forms of this species actually show different behaviors that also mimic those of their models.

The two models for *Hypolimnas* are from the genus *Amauris,* which belongs to the same distasteful danaid family as the monarch. *Amauris* means darkened, and you can see in the figure that these butterflies are quite dark. Predators in East Africa have therefore learned to avoid this dark pattern, rather than the orange-and-black pattern common among other danaids. The model on the left is darker than the other, and it tends to prefer shady places. It flies only briefly and in bursts, and rests on the undersides of leaves. The one on the right prefers sunnier areas, flies more slowly and for longer periods, and rests on the top sides of leaves.

The two mimetic forms of *H. dubius* mimic these two different behaviors. Indeed, each type goes so far as to emerge from its pupal case at the same time of the day as its model. The initial misclassification of these two forms of *Hypolimnas* as separate species is therefore not surprising.

H. dubius can also track other models with great precision. In East Africa it is also found in two forms, different in appearance from either of the West African forms. One of these mimics a different race of one of the model species of *Amauris* found in West Africa, and the other mimics a completely different species of *Amauris.*

In spite of their divergent appearance and habits, the two forms of *Hypolimnas* are without a doubt members of the same species. Males and females of the two forms have been observed to mate in the wild. Thorough genetic investigations have not been possible with this species, however, because they cannot be persuaded to mate in the laboratory. Nonetheless, the ratios of the two types seen in broods raised from females caught in the wild have shown that the form at the left of figure 7.3 is dominant to the form at the right.

This means that this whole complex of appearance and behavior is controlled by two alleles at a single genetic locus (location on the chromosome). Butterflies homozygous or heterozygous for the dominant allele at this locus will look and behave like the left form in the figure, while butterflies homozygous for the recessive allele will look and behave like the right-hand form.

This simple genetic system allows the mimic a great deal of flexibility. The two model *Amauris* butterflies are found in different proportions in different places. The frequencies of the two alleles in the various mimic populations reflect quite accurately the local frequencies of the two models. It is of course much easier to change allele frequencies in a population than it is to produce a whole new genetic pattern. By imitating two different models, *H. dubius* is protected against changes in model frequency and even against the disappearance of one of the models.

Of course, this flexibility comes at a cost. If models change in frequency, then the mimic form that is suddenly too frequent will be eaten in larger numbers. In the fecund and exuberant world of the tropics, this is probably not much of a problem.

But the genetics of the system does pose a problem—at least to geneticists. The fact that this whole set of characters is governed by what appears to be a single gene is quite startling. Indeed, it appears to go counter to everything we have learned about genetics over many decades of research—that one gene is responsible for one character. A gene might determine the synthesis of a single enzyme, so that even if there is more than one effect on the organism's phenotype, these effects can all be traced to the presence or absence of that enzyme. If more complicated genetic characters are investigated, they are almost invariably found to be controlled by the action of many genes at many loci, or locations, on the chromosomes, a good percentage of which have alternative alleles.

But consider the two mimetic forms of *Hypolimnas*. The color of the hind wings is white in one, yellow in the other. The beautiful fringing line of white spots that runs along the margins of the wings of one form is completely missing in the other. Other white spots have appeared or disappeared, changed shape or changed position in the two forms. In addition to these obvious pattern differences, there are the behavioral and developmental differences that we mentioned earlier. How can these two alternative complexes of very different characteristics be the work of a single gene with two alternative alleles?

They must be. *H. dubius* simply cannot afford to have its two different forms controlled by alleles of many different genes. If they were, and these genes were inherited independently, any wing pattern and behavioral differences would quickly break down into a bewildering variety of types. After a generation or two, few if any of the butterflies would resemble their models sufficiently closely to be effective mimics.

It happens that there is an answer to this apparent contradiction, and it is one that occurred to the first geneticists working on mimicry. The answer is that collections of different genes can be passed from one generation to the next as a unit, provided that they are right next to each other on the chromosome—*closely linked,* is the technical term. It seems quite certain that the differences between the two morphological types in *Hypolimnas* are due to the effects of a cluster of contiguous genes inherited as a unit. Such a cluster has been called a *supergene complex.* As a unit, it can affect more of the organism's properties than can any of the individual genes that make it up.

Hypolimnas provides one of the clearest demonstrations of the existence of such a supergene complex. We can already infer a good deal about how the complex might have evolved.

First, not all butterflies have such complexes. A great deal of genetics has been done on the South American *Heliconius* species that form such a conspicuous part of various Müllerian rings. They show no detectable sign of mimicry supergene complexes.

If you capture two butterflies of the same species, say *Heliconius erato*, from different areas, they will look very different because they will belong to different Müllerian rings. Mate the two together, and they will produce a great variety of different types in the next generation, few of which resemble either parent. This shows that the genes controlling the pattern are inherited independently, rather than clinging together in a supergene. It also shows that these types almost never meet each other in nature, because if they did, their precious Müllerian resemblance would break down.

This makes sense. If a heliconid matches a Müllerian pattern in its area well, then of course there is no need for it to match other patterns in the same area. Because so many species take part in a Müllerian ring, it is not likely that the ring will disappear if one or more species disappear. The heliconids are safe. They have no necessity to organize their mimicry genes into supergenes.

The second thing we can infer about these supergenes is that the *Hypolimnas* complex must have evolved as a result of the movement of genes from one part of the genome to another. Genes that are scattered widely on the chromosomes of the heliconids have been gathered together in the genome of *Hypolimnas*, perhaps through the agency of transposable elements.

How long did this process take? We cannot tell, but it will soon be possible to use the new tools of molecular biology to see whether genes homologous to those that are grouped in *Hypolimnas* are scattered around the chromosomes in nonmimetic nymphalids.

Indeed, it should also soon be possible to make an even more interesting comparison, one between *Hypolimnas dubius* and some of its closer relatives such as *H. dinarcha* that mimic only one model species rather than more than one. Are the morphology and behavior genes still scattered in these nonpolymorphic mimics? If so, then the supergene must have evolved very recently indeed.

I will stick my neck out and predict that when it is possible to do these experiments, that is indeed what we will find. We will see in a moment another reason for supposing that supergenes can evolve quickly.

The third thing we can infer about supergenes has to do with their structure. Batesian mimics have to be able to evolve quickly to keep up with their models, which (unlike the situation in Müllerian rings) are always trying to evolve away from them. As I have suggested, the situation the model faces is rather like the one facing the designers of new fashions, who must always be one step ahead of a horde of imitators. This ability to evolve quickly must be built into the genes that dictate pattern and behavior in both the models and the mimics. The easiest place to look for this enhanced evolutionary ability will be in the mimicry supergenes where all these genes have been brought together.

It seems very likely that once these complexes are examined in detail, they will be found to have a specific internal organization that lends itself to the production of new morphological types through recombination and mutation. *Hypolimnas* butterflies on the west coast of Africa are different from those on the east coast. The changes must have occurred within the complex. The process of crossing over might possibly produce useful variants in butterflies venturing into areas where there are different models, but any really new pattern would have to arise by mutation.

The really fascinating question concerns the degree to which the organization of the cluster lends itself to the production of new mutations. I would predict that when, as is inevitable, this cluster of genes is cloned and sequenced, it will be found to have as superbly functional an arrangement as do the clusters of immunoglobulin genes in man.

Look again at figure 7.3. Large changes in the pattern of a butterfly's wing can be brought about by the addition or removal of opaque white pigment in certain areas. In various butterfly families, white color in the scales of the wing is the result of the accumulation in the cells of different pigments. If such a pigment appears in the scales, it can mask the presence of other pigments. Inactivation of a gene for pigment production by the insertion of a transposable element could make the underlying pattern beneath a white spot suddenly appear. Reactivation by the removal of the element could make it disappear again.

Behavior, too, might be determined by a finite number of genetic states. Butterflies are sluggish or active, shade- or light-loving, and so on. It is possible to imagine mimics switching back and forth among various behavioral states as, in the course of evolution, the model alters its behavior to move away from the mimic or one model replaces another.

Most mutations simply cause damage, even if they occur in a well-ordered toolbox like the *Hypolimnas* supergene complex. But it seems likely that, while the complex itself has been evolving, mutagenic agents have been selected to settle in certain parts of the complex and make specific changes there. If the model species were to change dramatically, a combination of a highly structured supergene complex and these mutational agents would aid the swift evolution of a matching mimetic type.

Certain combinations of pattern and behavior must appear again and again in the course of the evolution of the models. Just as we saw with the antibody genes, the mimicry genes must by this time have become expert at changing in specific ways, so as to match the models quickly. But, just as with the antibody genes, the mimicry complex should be able to evolve swiftly even if a very different model appears on the scene. This must happen often enough to keep the evolutionary response of the mimics flexible—to keep them on their evolutionary toes, so to speak.

This is all highly speculative, because it has not been possible to investigate the supergene complex of *Hypolimnas* through genetic crosses. But more secrets of the supergenes have been found in other mimicking butterflies.

Papilio dardanus (male)   Papilio dardanus (female)   Amauris echeria

Figure 7.4   The butterfly at the left of the picture is the male of *Papilio dardanus*, showing the characteristic swallow tails on the hind wings. The middle butterfly is a female of the same species from the same area. The third is the model that this particular female mimics, *Amauris echeria*. Although the mimetic female is larger than the model, the wing patterns and the colors (though not visible in this photograph) are a remarkably good match. (From plate 9 of E. B. Ford, *Ecological genetics* [London: Chapman and Hall, 1975].)

TROY'S ANCESTOR

Of all the dazzling butterflies found in the savannahs and rain forest glades of sub-Saharan Africa, none is more brilliant than *Papilio dardanus*. It is over four inches across, a great prize for collectors.

P. dardanus belongs to the family Papilionidae, and its family and genus names both mean simply butterfly. Along with most of the other members of its family, it has long swallow-like tails on its hind wing and a striking yellow-and-black pattern. Some dim impression of this can be gathered from its picture in figure 7.4.

It is named after Dardanus, the mythical founder of Troy. He was the son of Electra, one of the Pleiades, the daughters of Atlas. This suggests, although rather remotely, the butterfly's African origin, and adds a touch of the exotic and mysterious. These are properties that this species certainly has in abundance.

mysterious. These are properties that this species certainly has in abundance.

P. dardanus is even more widely distributed through Africa than *Hypolimnas*. It is found throughout the sub-Saharan regions, and in the east as far north as Ethiopia. It is also found on the island of Madagascar, which lies off the east coast of Africa and is a kind of evolutionary museum filled with species of animals and plants found nowhere else.

In common with other swallowtails, *Papilio* is a strong flyer. The males are more active than the females, which sit for long periods in exposed positions during the mating season. It may be that this difference in behavior helps to explain a remarkable pattern that became apparent as specimens of this butterfly began to be collected during the exploration of Africa a century ago. While the males look quite similar wherever they are collected, the females are very diverse. Many of them are close mimics of whichever distasteful model species happens to be present in that particular area. In many cases, the females found in a particular region come in different forms, mimics of two or more distasteful model species. Most, though not all, of the species mimicked belong to the poisonous danaids. These mimicries must provide the female *Papilio* with a strong advantage during her long and dangerous wait for suitable males.

The mimic females are smaller than the very large *Papilio* males, their color and wing patterns are completely different, and they have lost the distinctive swallow tails on the hind wings. As a result, the females of *Papilio* are just as effective mimics as are *Hypolimnas*, and indeed some of the species that they mimic are the same as those imitated by *Hypolimnas*.

There are two exceptions to this pattern. The first is found on Madagascar and in northern Ethiopia, where distasteful models are rare. Here, both males and females share the distinctive swallowtail shape and conspicuous black-and-yellow markings of the nonmimetic *Papilio*. The second exception is found throughout the range of this species on the African mainland. Almost everywhere, there are a certain number of females that do not resemble the males but also do not have a close resemblance to distasteful model species. This is particularly marked in the mountain regions east of Lake Victoria, where model species are also rare. Here the females, in addition to showing a variety of patterns and colors, sometimes also show traces of tails on the hind wings and patches of the typical yellow pigment that is prominent on the males.

This complex pattern of mimetic and nonmimetic forms can be interpreted in a number of ways. For instance, it has been suggested that the race found on Madagascar and in Ethiopia might represent the ancestral condition of the species: both the males and the females resemble each other and there is no mimicry. But an equally valid possibility is that *P. dardanus* on the mainland of Africa has always been mimetic, and that its ancestors were mimics too. The Madagascan and Ethiopian races would then be the result of fairly recent relaxation of selection against the nonmimetic type. This might be happening in the region around Lake Victoria as well, but the process is slowed by the immigration of mimetic females from surrounding areas where they are common.

Of the two possibilities, I suspect that the second is more likely. At the very least, it would have given the complex genetics of this species plenty of time to evolve, since it is always possible that the mimicking abilities of *P. dardanus* actually evolved, not in this species, but in its ancestors.

Regardless of when or how it developed, there is a remarkable supergene complex in this species, and it exists in a variety of forms. The integrity of these various forms is maintained by strong selection, and they tend to break down through crossing over when selection is weaker.

The situation cries out for genetic analysis. Luckily, it has been possible to do some. During the 1950s and 1960s, Cyril Clarke and Philip Sheppard were able to collect and have sent to them *Papilio* from many parts of Africa. They worked at the University of Liverpool, an unlikely place to raise tropical insects. But the butterflies could survive in heated greenhouses; and their caterpillars thrived on a plant (actually a Mexican import) found in local gardens.

One of the great advantages of this species, unlike *Hypolimnas*, is that it can be hand-mated. If the abdomens of a male and female are pressed together by the investigator, nature can be persuaded to take its course. This allows matings to be made between various races of butterflies that would normally not meet.

The first thing that Clarke and Sheppard discovered was that the presence or absence of a tail on the hind wing was due to homozygosity for one of two alleles at a single genetic locus. Butterflies homozygous for one allele had tails, regardless of whether they were male or female. Butterflies homozygous for the other allele had tails if they were males and did not if they were females.

Heterozygotes are found in nature only in those mountainous areas near Lake Victoria where the Ethiopian and the southern races overlap. Heterozygous females have small tails of variable size. The Madagascan and Ethiopian populations are homozygous for the allele that produces tails in both males and females.

So that part was clear enough, and surprisingly simple. Because all the mimics are tailless, all the races of *P. dardanus* with mimic forms are homozygous for the allele that prevents the formation of tails in females but not in males. The inheritance of this character has nothing to do with a supergene complex.

Clarke and Sheppard found the real supergene complex on another chromosome. This consisted of a series of "alleles," each actually a complex of closely linked genes, that controlled color, pattern, and probably many other things. No butterfly could have more than two of these alleles at once, of course, but there were many different alleles in the species as a whole. Some were dominant, some recessive. Still others were dominant when heterozygous with some alleles of the series and recessive when heterozygous with others.

The supergenes were not the whole story. When Clarke and Sheppard crossed strains from very different places—Madagascar and South Africa, for instance—they found that supergene alleles that used to be dominant and clear-cut in their effects no longer worked properly in the new genetic background. The mimics no longer matched their models well. It turned out that the supergene alleles had coevolved along with many other allelic differences on other chromosomes to produce a phenotypic match with the local models. The supergene always gave a good match to the model in the race in which it had been selected, but it gave a poor match in other racial genetic backgrounds.

Again, we do not know how old or how widely shared this supergene complex is. *Papilio* belongs to a butterfly family that is different from *Hypolimnas*, and it may be that its supergene evolved independently. Time, and the inevitable activites of the people who sequence DNA, will tell.

HOW THE MIMICRY SUPERGENE COMPLEXES FACILITATE EVOLUTION

The mimicry systems of *Hypolimnas* and *Papilio*, in spite of their differences, have many things in common. In each case, the primary matching to the various model species is accomplished by a supergene complex. In *Hypolimnas* the complex has its effect in both sexes, while in *Papilio* it affects only the females, but this is a relatively trivial difference. *Papilio* males also carry the supergene complex, but it is simply not expressed, probably due to the effects of a sex-limited repressor gene present in this species that is carried by both sexes but expressed only in males.

Further, in both of these mimic species more than one supergene allele can exist in a population simultaneously. The usual situation in temperate regions is for a mimic species to imitate only one model, but the world of African butterflies is so complex and rich in potential models that it is possible and often perhaps necessary to mimic more than one. No one model may be common enough to provide the protection needed to maintain the mimic population in substantial numbers.

How long does it take to alter these supergene complexes and produce new ones? This is central to the question of how easy it is for evolutionary change to take place. Our great difficulty in trying to answer it is that we do not know how much time we should allow. There is no fossil record for these African butterflies, and in any case wing pattern and color would not be preserved as fossils. But three evolutionary scenarios immediately come to mind.

First, it is possible that all these species have been around for many millions of years, during which time this complex mimicry pattern slowly evolved. This seems rather unlikely. The African tropics are indeed very old, but in general, tropical species do not persist any longer than temperate ones.

A second possibility is that the supergene alleles are actually older than the mimicking species themselves. Genetic polymorphisms can indeed survive the genetic upheavals associated with the founding of new species. This seems to have happened with the ABO blood group polymorphism, which is present in many of our primate relatives as well as ourselves. But this hypothesis implies that the ancestors of both the models and the mimics looked very like them even though they were different species. The mimic-model relationship we see at the present time would be required to persist even during the genetic changes that led to the formation of new species. This also seems unlikely because the color and pattern of butterflies is the most genetically malleable thing about them. Closely related species are often very different in appearance.

The third and most likely possibility is that mimics and models are in a continuous state of flux. In addition to the long-term evolutionary changes as Batesian models try to escape from Batesian mimics, there are seasonal fluctuations as well. Model species can become more or less frequent with time, and the geographic range over which they are found can alter. Surveys have shown that model species often do fluctuate in numbers from year to year, and that mimetic species increase and decrease in numbers in synchrony with these fluctuations. It must be a constant race for the mimetic species to keep up with these rapidly changing conditions.

But let me emphasize again that even if the environment and the mimic-model situation are constantly changing, there are not, of course, an infinite number of model patterns. Particular patterns or parts of patterns that the mimic must match will appear time and again in different models as the African butterfly community evolves.

How does this affect supergene evolution? To understand this, we must separate the evolution of the mimic supergenes and the evolution of the mimics themselves. Butterfly models evolve, change in numbers, and migrate to new areas. Butterfly mimics must also evolve to keep track of these events, and they do so through changes in their supergenes. And, at a much slower pace, the supergenes themselves evolve.

The Louis Vuitton handbags and their cheaper mimics provide a good analogy. A proliferation of mimics will eventually force the original company to change its product. This requires the expenditure of much effort in design, marketing research, and advertising, so it makes such changes rather seldom and at great expense. And it makes them at considerable risk, for the new product may be a failure. Consumers may not care for the new model. This is the equivalent of a dramatic change in the appearance of a model butterfly, which is likely only rarely to occur successfully—that is, to spread through the population and replace the older models. There is a cost associated with this change also, for predators must be trained to avoid the new model pattern. And any reorganization of the genes probably has some cost in terms of reduced fitness.

But the mimic hand-bag factory must merely copy a successful product, so it can be organized entirely for production and need devote nothing to design or advertising. And it must do so as efficiently as possible, because speed counts. This is the equivalent of the evolution of a new mimic form. If the mimic population can produce a new mimic form quickly, it will be able to track the model population without great risk to itself.

Over time, the ways in which both the model and the mimic hand-bag factories are organized will also change. But because speed is of the essence, the organization of the mimic factory will change more quickly than that of the model factory. I have not done a study of this, but I suspect that mimic factories in Hong Kong go extinct fairly often. The best-organized factories, those most quickly able to bring their mimics on the market, will survive.

A similar dynamic must operate in populations of Batesian mimics. When we examine the organization of supergenes in such populations, we should be alert to the possibility that they have evolved not just to keep genes together but to facilitate their further evolution. As we saw with the antibody complexes, the alteration and regrouping of genes within the mimicry supergene may be facilitated by viruses and transposable elements, particularly if such elements have co-evolved with the supergene. This should greatly increase the likelihood of the appearance of new and potentially useful mimic types.

It is possible to see how this evolution of the organization of the supergene itself can proceed fairly swiftly, by individual rather than by species selection. You will remember that in both *Hypolimnas* and *Papilio* different supergene alleles can coexist in a single population. While we do not yet know how these supergenes are organized, it does not require too much of a leap of faith to suppose that different alleles have different internal organizations. Variation in internal organization *within a species* would in turn mean that some members of the species would have a higher probability of producing adaptively useful offspring than others. Thus, we do not need to wait for extinction events to weed out less well organized supergenes and preserve the ones that are better organized. This can happen swiftly, as the pressures of the evolution of mimicry demand.

8

The Evolutionary
Orient Express

In some cases, regeneration is "heteromorphous": in [the stick insect] *Carausius*, if the antenna is cut through at the level of the flagellum, it regenerates as an antenna; if it is amputated through one of the two basal segments it commonly regenerates as a leg, complete with tarsal claws.
                                    —V. B. WIGGLESWORTH, *Insect Physiology*

It is difficult for us to imagine any group of higher animals more utterly different from ourselves than the insects. We have all watched nature films in which enormously magnified insects unfeelingly dismember their prey. Their glittering multifaceted eyes stare at the camera while their complex mouthparts work busily, munching through still-struggling victims. We can empathize with our closer relatives the lions, who at least seem to enjoy their bloody work. But when the female mantis bites the head off its mate in order to release its copulatory reflex, it does so at the behest of an instinct that seems to have nothing to do with love, hate, or anything else to which we can remotely relate.

Of course, we must not fall into the trap of being anthropomorphic. Insects are naturally expressionless, since they wear their skeletons on the outside. One of the terrors of medieval battle was the sight of a fully armored man, iron-visaged and machinelike, hewing his way through a tide of enemies. Inside the armor he was probably expressing all kinds of emotions. Somehow, though, I doubt that insects, inside their armor, are expressing anything that we could classify as an emotion at all.

How have we come to share the planet with such complex, highly developed, and yet totally alien creatures? The insects have obviously taken a very different evolutionary path from ourselves, but it is only just becoming apparent why this is so. We can now see that the organization of insect genes has had an immense and continuing impact on the course of their evolution, and this explains why it has been so different from our own.

The best way to follow the divergent insect and vertebrate paths is to imagine ourselves traveling to a very remote evolutionary time, the earliest for which we have good fossils of multicellular life. Let us adjust the controls of our time machine to take us back 600 million years. Feel free to imagine suitable special effects. At the end of our journey, we suddenly find ourselves on a grim-looking shore.

This world is utterly different from anything we know. Life is still essentially confined to the sea, which spans most of the globe. The shore is in fact the edge of a vast, empty supercontinent that geologists have christened Pangaea I. Aside from a few scattered islands, it is the only piece of land on the planet.

Some tens of millions of years in the future this supercontinent is scheduled to begin a slow breakup into smaller continent-sized landmasses. These will not be the continents we are familiar with, however. We have gone back too far in time for that. The continents that result from the breakup of Pangaea I will go their separate ways for a few hundred million years, then coalesce again to become a second supercontinent, Pangaea II. It is Pangaea II that will break up into the beginnings of the present-day continents at the start of the

Age of Reptiles some four hundred million years down the road from the time of our visit.

The lifeless interior of Pangaea I is even more remote from the moderating influence of the ocean than is central Asia today. Much of the time it is covered with immense glaciers. The erosion that results from their seasonal melting is unchecked by plant life, so that mountains wear away rapidly and the rivers are choked with silt. Huge estuaries form at the mouths of the rivers.

In the oceans, life has already been evolving very slowly for perhaps three billion years, but now things are beginning to speed up. Photosynthetic bacteria, aided more recently by one-celled algae, have been producing oxygen for about half of this time, slowly changing the composition of the very atmosphere itself.* The oxygen has dissolved in the water in ever-increasing amounts, and not long before the time of our visit this concentration has reached the point at which multicellular life can be supported.

The great estuaries are ideal places for the further evolution of these early multicellular organisms. We have surprisingly good records of some of the soft-bodied creatures that lived there. Fossils from this remote time have been preserved in coarse sandstone from the Ediacara Hills near Adelaide in the state of South Australia. The fossils, found in the 1940s, were thought at first to be unique. But soon similar fossils were discovered in many other parts of the world, notably southwestern Africa, the English Midlands, and various parts of the Soviet Union. As a result, we know some of what you might see while tramping across the tidal mud flats at the mouth of one of the ancient rivers.

The countryside looks much like the barren saline estuary of the Colorado River. The ancient river that formed the es-

---

* Recent evidence suggests that oxygen-producing organisms evolved as long ago as 3.3 billion years, 2.7 billion years before Pangaea I. But free iron in the crust and the oceans absorbed any oxygen produced until about 2 billion years ago.

tuary, however, is far larger than the Colorado and flows swiftly all the way to the sea. (The Colorado, depleted by irrigation, disappears into the sand long before it reaches the Gulf of California.) The river is heavily laden with silt and foams its way between erosion-gouged banks until it finally spreads out across tidal flats more extensive than anything we see in our present world.

At the time of our visit the tide is low, and the great alluvial fan of the estuary, made up of mud and coarse brown sand, stretches endlessly away from the coast. The ocean is simply a glimmer at the edge of the world, where the horizon is as sharp as a knife in the clean air of the planet's youth.

Aside from their extent, the tidal flats themselves are not very different from some of those in our present world. The receding tide has left a few flat creatures up to a foot across drying in the sun. They vaguely resemble jellyfish, but they have an oddly quilted texture. You turn over some small rocks half-buried in the sand and find small segmented worms clinging to them, occupying the same niche as the segmented marine worms we find on the bottoms of intertidal rocks to-day. Deeper digging, however, reveals nothing. None of these animals have begun to exploit the vertical component of their environment.

In tide pools there are a few creatures looking like old-fashioned quill pens, with plumes that appear to have been left behind by some bird, their tips buried in the sand. They are sea pens, and they have come down to us essentially as they were six hundred million years ago.

Another organism that would be familiar to present-day zoologists is represented by scattered ovoids of what look like clear plastic left behind by the tide. These are the floats and attached sails of siphonophores, little animals that literally sail the ocean. Their close relatives today are *Velella*, the tiny "sailors" that are driven by the wind in great circles on the open ocean and are sometimes washed ashore in un-counted numbers. Siphonophores actually represent an early stage in the evolution of multicellular life; they consist of a colony of specialized organisms that have various functions and cannot survive separately. The floats that crunch under your feet are only a small indication of the life that teems in these ancient oceans.

As you poke about, you see creatures that would puzzle even a trained zoologist. There are, for example, little tripartite jellylike organisms that resemble nothing alive today. This is not surprising, for all their descendants died out at some point during the intervening 600 million years.

Up near the shore are bumps of rock, rather like smooth corals. These are formed by mats of algae and bacteria growing and building up in successive layers. They provide a connection with a much earlier time, when these mounds, called stromatolites, were the most elaborate products of living organisms on the planet. At the present time a few stromatolites are still being constructed in tidal pools in western Australia and Baja California, perhaps by similar organisms. Through much of the history of life on the planet, down to shortly before the time of our visit, these stromatolites were very widespread and were perhaps the only sign of life that a time-traveling tourist lacking a microscope might see.

Your walk also reveals creatures for which we have no fossil record. There are certainly algae, perhaps in the shape of tangles of simple green filaments, perhaps forming more elaborate structures. Attached to the bottom of the tide pools, there might be some little filter-feeding animals with bulbous bodies, extending their gills widely to take in as much oxygen as they can. Zoologists have speculated that such filter feeders were our remote ancestors. Their traces have not been found in strata dating from this time, but they were probably present. (None of these creatures has a shell or a skeleton. These will come fifty million years later and will trigger the explosive radiation of multicellular organisms in the Cambrian.)

It is not, to the uninitiated, a particularly exciting walk. For one thing, it is probably rather uncomfortable. The oxygen-poor air quickly leaves you short of breath as if you were slogging around at a high altitude instead of at sea level. And it is cold. We always imagine the primitive earth as a hot and steamy place, but instead a cold wind from the distant gla-ciers ruffles the standing pools of water on the tidal flats. If you venture too far out, you might be caught by the swift return of the tide. The moon is close, and the tides are cor-respondingly huge. This explains the enormous extent of the estuary. Traces of these huge tides can be seen at the present time, in the form of rock layers called *varves* preserved in the dry southern Australian desert.

You will probably be glad after your hike to get back to the dry warmth of your time machine, and will feel a good deal of relief during the swift journey home. The Precambrian earth is an alien and forbidding place. But many biologists would cheerfully sell their souls for a few hours on those cold and dismal flats. For this is where much happened that shaped the whole subsequent course of history on the planet.

Even at this remote date, a great evolutionary split had already separated the ancestors of the insects from those of the vertebrates. We have no fossil record of this split, and the degree of molecular divergence between these two groups is so great that we have only an approximate idea of when it might have occurred.

Indeed, even were we to find their fossils, we would prob-ably not even recognize these creatures, the remote ancestors of both ourselves and the butterflies. They would likely ap-pear superficially to have nothing in common with either group. They were certainly quite different from the seg-mented worms and filter feeders we have just found on our imaginary trip to the mud flats: these had already branched long before in a variety of different evolutionary directions.

During this Precambrian time, now commonly called the *Ediacaran,* and the Cambrian period that immediately fol-lowed it, all the phyla of animals evolved. This was a remarkable series of events, leading us to ask why so much evolutionary change took place just then.

There was probably more than one reason for this adaptive explosion. Most important perhaps was the presence for the first time of appreciable amounts of free oxygen in the atmosphere, enough to allow briskly metabolizing multicellular animals to evolve. This was almost certainly the trigger that led to the appearance of the primitive multicellular creatures that we saw in the mud flats.

Later, the supercontinent Pangaea I began to break up. Shallow seas and substantial areas of continental shelf appeared, which provided many new ecological niches. This breakup, coupled with the appearance of ozone in the upper atmosphere, also probably did much to moderate the world's climate. Warmer shallow waters led to complex associations of species of the kind that we see in today's tropical reefs. Complex ecosystems generated new ecological niches, leading to the appearance of different and ever more elaborate organisms. These largely replaced the simple Ediacaran creatures, and many of them were so successful that their descendants still live among us.

The Cambrian adaptive explosion was remarkable, and indeed unique, because of the astonishing proliferation of different *types* of animals. Mammals went through the same kind of adaptive expolosion sixty million years ago, but at the end of it all of them, from bats to whales, were still recognizably warm-blooded milk-producers. But the explosion that started with the Ediacaran and was essentially over by the end of the Cambrian, spanning some 200 million years, led at first to the strange creatures we saw on the mud flats and later to all the types of animals that inhabit our world. These included creatures as varied as worms, chordates, echinoderms, and molluscs, along with many other exotic creatures that have left no descendants. Why this astounding burst of creativity on the part of evolution?

The paleontologist James Valentine has suggested that because the animals of that time were relatively uncomplicated, they were better able to survive dramatic reorganizations of their body plans as a result of mutations. This was, perhaps, a time when hopeful monsters played a much larger evolutionary role than they do today.

There is an automotive analogy to Valentine's supposition. At the beginning of this century, cars were powered by gasoline, kerosene, even coal and wood. Their engines were internal combustion or steam. Manufacturers could easily switch back and forth among these various options because most cars were collections of machinery bolted together without much regard for efficiency or aesthetics. (The Model T Ford could be converted to burn alcohol by simply moving a lever on the carburetor.) These early cars were noisy, unreliable, wasteful of energy, and fearfully polluting. They survived in spite of this because they were such an improvement over the horse and carriage that they replaced.

Now consider the common ancestor of the vertebrates and arthropods. It can be thought of as a kind of evolutionary Model T or Stanley Steamer. In that simpler world it was not constrained by an elaborate structure but was quite capable of evolving in a number of directions. It must have had a very simple genetic organization as well. Mutations and rearrangements of this simple collection of genes gave rise to a great variety of descendants. Most of these have probably disappeared without a trace, but the survivors include ourselves and the insects.

Let us see what kinds of selective pressures might have produced the split that eventually led to ourselves and to the insects. The split happened very early, so early that the long trail of subsequent evolution has had plenty of time to produce very different kinds of genetic organizations in the insects and in the mammals.

We can begin with our own story. By the time of our visit via time machine to the Precambrian estuary, our ancestors had probably evolved from this primitive progenitor into ses-

sile arm-feeders, perhaps rather like soft-bodied goose bar-
nacles in appearance. Only the adults were sessile. The larvae
were able to move about, but once they settled down, they
made a permanent commitment. If the area was satisfactory
they could devote themselves to their two adult occupations,
filter feeding and producing young.

There was therefore strong selection acting on these larvae
for the ability to swim rapidly through the water. This would
have two advantages: they could swim longer distances, increasing their chances of finding a suitable settling place, and
they could also escape predators.

The larvae of these filter feeders evolved a unique mechanism for swimming, a stiffening rod of cartilage that started
to develop near the anus and when fully mature extended
most of the length of the body. This is called a _notochord_
(literally, back stiffener), and it still appears briefly in our own
embryonic development before it is replaced by the stronger
and more flexible vertebral column. The notochord gave the
larvae, when they needed it, a little backbone. Combined
with paired segmented muscles that developed along the
body, it enabled them to wriggle their way through the water.

It seems likely that the free-swimming larvae in one branch
of this early family tree matured sexually even before they
settled down to their sessile adult existence. Soon, the sessile
adult stage was lost in this lineage, since the entire life cycle
could be completed without it. These free-swimming creatures, able to remain mobile throughout their whole lives,
could move into many new ecological niches. This branch
developed into the true _chordates,_ as organisms with a notochord are collectively called. And it is this branch that eventually led to us, for we are also chordates, in a subgroup
called the vertebrates.

We have a good idea that this is the way it happened because there are some marine relatives of ours, the tunicates,
that exhibit this very pattern. The free-swimming larva have
a notochord, vertebrate-like striated muscles, eyes, a balanc-

ing organ, and other chordatelike structures. The adult tunicate, however, is a sessile filter feeder. It has lost all the chordate features, except that its muscles are like those of vertebrate smooth muscles. It must be admitted, however, that although this evolutionary story makes sense, the fossil record is absolutely blank on the subject.

Selection acted very differently in the lineage that led to the insects. You will recall that when you dug into the mud of the Precambrian tidal flat, you found no creatures. This is because very few organisms from that time were able to burrow, and of those, none was able to burrow deeply.

But the primitive worms that you found clinging to small rocks on the flats were able to do a little burrowing. This was not because they had evolved a notochord, but because their bodies had become divided into compartments called segments. And it is these organisms that eventually gave rise to the insects. By the time of our visit to the tidal flats, some had already evolved paired rows of jointed legs, two to each segment.

A completely unsegmented, soft-bodied organism would have great difficulty moving through a substrate that resisted it, since it could not force a path. Various ways have been found around this problem. Amoebas that live in the soil flow through any tiny opening and re-form on the other side. But they cannot grow large, because they consist of only a single cell. Nematodes, extremely common minute roundworms, also live in huge numbers in soil. They can move between the soil particles both because they are small and because they have evolved a tough outer cuticle that gives their bodies stiffness and strength. Although they are multicellular, they also cannot grow large, because if they did their bodies would be too stiff for flexible movement.

The segmented worms have solved the burrowing problem by means of compartmentalization. Their bodies consist of a series of compartments separated by partitions, rather like the Orient Express. This gives both flexibility and stiffness.

Each compartment in the simplest segmented worms is constructed on the same basic plan, with internal structures and appendages confined to that compartment. Each compartment of a segmented marine worm, for example, has a pair of leglike appendages, a pair of excretory organs that act as simple kidneys, a nest of tiny tubes leading to the outside of the body that serve in respiration, a ganglion for nervous control of the segment and communication with adjacent segments, and so on.

The body plan is not, of course, simply a series of precisely replicated compartments. Even the simplest present-day worms have evolved specialized compartments and exhibit connections from compartment to compartment. In particular, the gut and the circulatory and nervous systems cross the segment boundaries. Nonetheless, the functions of each compartment are remarkably autonomous.

Some segmentation is seen in the chordates too, and this is particularly marked during early development. One striking example is seen in a particular set of paired blood vessels—which in our water-breathing ancestors formed a series of gills—and their supporting structures, the gill arches. We still have these gill arches during our early development, though they no longer function as breathing apparatus. Later in our fetal life some of the arches undergo a remarkable series of transformations to become such varied structures as the bones of the jaw and of the middle ear. We show other traces of this early segmentation in our abdominal muscles and in our paired cranial nerves and the paired nerves that issue from the spinal column through openings in our vertebrae. But in most adult chordates, the confining effects imposed on development by segmentation have largely disappeared.

In the most primitive chordates segmentation is still pronounced. One of these, a small slipper-shaped transparent creature called a lancelet, has a notochord but also shows a clear-cut division of the body into segments. Today, the lancelet lives both in tidal sandy flats and in deeper water. There may have been lancelets in those Precambrian estuaries, but no sign of them has yet been found in the fossil record.

The big difference between the vertebrates and the arthropods is not the fact of segmentation itself, but rather the degree to which segmentation persists during development and dictates the body plan. We vertebrates have escaped from the tyranny of segmentation to produce organs and structures that cross the segment boundaries. The insects, by and large, cannot. The structures in each compartment remain trapped within its confines. It is as if the compartments on the Orient Express had no connecting corridors.

As a consequence, the insects have been unable to grow very large. Internal skeletal support never evolved because it was not needed early on for stiffening the body. Instead, the external cuticle thickened and took up both a protective role and a role as support for muscle attachment. The muscles in insects are tiny, numerous, and attached to the insides of their exoskeletons at many points. This works well when they are small, although insect muscles are no stronger for their size than vertebrate muscles. But diffuse muscle attachment does not work well when size increases. Our own muscles, larger and much less numerous, attach at specific points to our skeletons and provide needed direction and thrust.

Breathing, too, is a problem with insects. They have not evolved true lungs. Their relatives, the spiders, do indeed have lunglike structures, but like insects, the spiders are limited in their size by the exoskeletal problem. In the insects, and in their aquatic relatives the crustaceans, oxygenated air or water is carried to the tissues through tiny tubes connected to the outside. Muscle action helps in this transfer, but the enormous and forceful exchange of oxygen-rich for oxygen-depleted air that is possible in a vertebrate is not possible in an insect.

So, the insects took the express train in a very different direction from ourselves. While the insect fossil record is rather sparse, it is apparent that insects have not changed dramatically in size or body plan at least since the end of the Carboniferous, some 300 million years ago. During the Carboniferous some were briefly able to grow larger, apparently because of increased levels of atmospheric oxygen produced by the immense quantity of plant life at that time. Fossil dragonflies with two-foot wingspans have been found in Carboniferous deposits. But beetles from that time looked very much like the beetles of today.

The early evolution of insects must have been very rapid and dramatic. Further dramatic changes certainly occurred at the time of the spread of flowering plants, partway through the Age of Reptiles, but the insect fossil record is too sparse for us to be sure of their extent. Yet, very early on, the insects hit evolutionary roadblocks of the kind that the vertebrates did not encounter.

We can now begin to see dimly what the nature of these roadblocks might be. They have to do with the way insect genes are arranged; they seem to have the effect of limiting certain evolutionary directions. If we can begin to understand how these gene arrangements constrain evolution, then perhaps we can also begin to understand how others might facilitate it.

THE FRUIT FLY AS A DEVELOPMENTAL TOOL

Drosophila melanogaster is a geneticist's dream. If you were given the task of designing from scratch the ideal multicellular organism for studying genetics and development, you would have to be supernaturally clever to come up with one that has all the handy features of the fruit fly.

The fruit fly has everything. Its life cycle is rapid, completing an entire generation in a little over ten days. There is no waiting for the next growing season in a fruit fly lab. It has a small number of chromosomes, with a small amount of DNA per cell—only about a tenth as much as we have. There are many clearly defined structures on its cuticle that can be examined for developmental mutations. It is easy to raise in the laboratory, and so far it has not been discovered by the animal rights activists.

The fly lays its eggs on the exposed flesh of soft fruit such as bananas, melons, or grapes, and the eggs hatch in twenty-four hours into little white grubs, or larvae. The larvae gorge themselves furiously for the next five days, increasing in size by a factor of more than a thousand. They can force their way through a soft substrate because, like the adult, they are divided into segments. Within these segments there develop other structures that will eventually give rise to the segments of the adult.

After this period of frantic gourmandizing, the larvae crawl out of their food and stop moving, and their outer cuticles become thick. The larval tissues are then broken down and transformed into the tissues of the adult, a process that takes about four days. This transformation is a profound one, and there is little resemblance between larvae and adults. The adults emerge from the old larval skin, their wings stretch, and their cuticle quickly hardens. Their preadolescence is brief: the females are ready to mate and lay eggs after about eight hours.

There are many other marvelous features of the fruit fly, and I will briefly list some of them.

While the adult cuticle is opaque and pigmented, the larval cuticle is translucent and can be made quite transparent by chemical treatment. If you have a Drosophila gene that you have cloned, you can label many copies of your cloned piece of DNA chemically and bathe a larva in it. Cells of the larva in which that gene is active make quantities of messenger RNA that can hybridize with your gene. Once the unhybridized DNA has been washed away, the parts of the larva in which that gene is active can be seen clearly.

Because the larvae must liquefy their food before they eat it, they have enormous salivary glands made up of grossly swollen cells. Everything about these cells is gigantic, including their chromosomes. By one of the happy coincidences of science, the giant chromosomes were discovered almost thirty years after the flies themselves entered the laboratory, just in time to clear up many genetic puzzles that had accumulated over that period.

The giant chromosomes of the salivary glands are one of the wonders of the natural world. They consist of ordinary chromosomes that have duplicated many times but that still cling together. This means that the flies have *made clones of all their own genes*, and arranged them all in a neat sequence like magazines bundled together by zip code. Each gene forms a band, and most of these are easily visible under the microscope. Figure 8.1 shows a set of these giant chromosomes, with each band, including many not visible in the picture,

---

Figure 8.1 (*overleaf*) A part of one of the giant chromosomes of *Drosophila melanogaster*. Each of the many bands on the chromosomes, including many that are not visible, represents one or a few genes. You would not be able to see these structures at all if the chromosome threads had not duplicated again and again, forming these abnormally thick sausagelike chromosomes. The clusters of black dots you see here and there represent the places where a cloned mobile element allied to *copia* has been hybridized to the chromosomes. The combination of numbers and letters shows the chromosomal locations of the genomic copies of this transposable element. (The single black dots scattered through the picture are due to background radioactivity.) Four of the five copies appearing in picture *A* have disappeared in the picture *B*. Picture *C* shows a subsequent generation in which new copies have been inserted into the same places! These events occur infrequently, but when they do, many transpositions occur simultaneously in a single line of flies. Nothing is known about what triggers these remarkable events. (From figure 4 of T. I. Gerasimova, L. V. Matjunina, L. J. Mizrokhi, and G. P. Georgiev, Successive transposition explosions in Drosophila melanogaster and reverse transpositions of mobile dispersed genetic elements, *EMBO J.* 4[1985]:3773–79.)

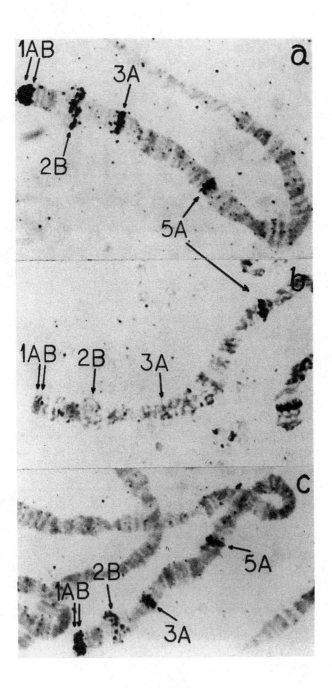

representing a gene cloned by the fly for the convenience of geneticists and molecular biologists.

The DNA-bath trick can be performed on the giant chromosomes, just as it can on the whole larva. You might have a piece of Drosophila DNA that you have isolated and perhaps cloned in a bacterium in order to make it pure, but you have no idea from what part of the Drosophila genome your DNA comes. The solution to this problem, as worked out some years ago by Mary Lou Pardue and Joseph Gall, is simple and elegant. Simply make your DNA highly radioactive or label it chemically and bathe the giant chromosomes in it. Your DNA will find its complementary half, thoughtfully cloned into many copies by the fly itself, on the chromosome. This elegant trick allows you to "light up" the region of the chromosome where your cloned piece of DNA lives—its address, so to speak.

Close examination of the giant chromosomes allows even small changes in the order of the genes to be seen. Geneticists have spent decades building up collections of flies in which pieces of the chromosomes have been deleted, inverted, or rearranged. The effects of such shufflings can now be explored in all parts of the Drosophila genome.

Tricks are also available for maintaining lethal genes in flies as a kind of permanent heterozygote, from which homozygotes can be produced whenever the geneticist wishes. This enables Drosophila workers to study genes that, when homozygous, disturb the fly's development so much that it dies as an embryo.

All these tools and more can be brought to bear on the fascinating problem of how these flies—and insects in general—evolved. As if to aid us in this, the flies have provided a remarkable series of mutants that look at first sight like hopeful monsters.

MOUCHES SCRAMBLÉES

A rakishly Latin diagrammatic representation of a fruit fly, adapted from a drawing by the Spanish geneticist Ginés Morata, is shown in figure 8.2. You can see clearly the segments of the thorax and the abdomen. The head also consists of a series of much-modified segments, but they are not shown here.

The best-studied segments are the three that make up the thorax, or middle section, of the fly, and the eight that make up the abdomen. This is because two remarkable collections of genes—supergenes that bear a superficial resemblance to the ones we examined in the butterflies of chapter 7—control their development. These supergenes are called the *Bithorax* and *Antennapedia complexes* (BX-C and ANT-C for short).

These complexes have fascinated developmental geneticists for decades because mutations in these genetic regions have large effects on the fly. Some of the most striking mutations have the property of allowing structures normally found in one segment to appear in other segments. Because of these effects, it can be shown that the major function of BX-C is to control the development of the hind half of the fly's body.

We can determine what BX-C does by the simple experiment of removing it and seeing what happens to the fly in its absence. This permits us to see the effects of other genes that are normally masked by BX-C. In a mutant in which BX-C has been deleted from the chromosomes, all the segments posterior to the middle thoracic segment begin to develop in the same way as that segment does.

You will see from figure 8.2 that the middle thoracic segment is the one that has the wings and the middle pair of legs. When BX-C is present, it is responsible for developmental changes that take the basic pattern of the middle thoracic segment and modify it. This results in the patterns seen in the last thoracic segment or the various segments of the

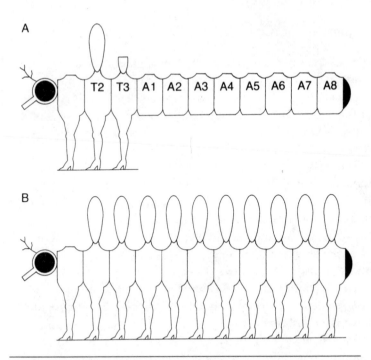

Figure 8.2 (*A*) A diagram, modified from a picture originally drawn by Ginés Morata, showing the various segments of a fruit fly's body. The segments of the head are difficult to understand, and are not shown. The important points to be seen from this figure are that the thoracic segments T1, T2, and T3 all have legs, but only the middle segment, T2, has wings. The eight abdominal segments A1 through A8 have neither legs nor wings in the wild-type fly. (*B*) A visualization of what would happen if a fly in which the BX-C has been deleted could reach adulthood. All the segments posterior to T2 would be similar to T2. (Modified from figure 1 of D. S. Hogness, H. D. Lipshitz, P. A. Beachy, R. B. Saint, M. Goldschmidt-Clermont, P. J. Harte, and S. L. Helfand, Regulation and products of the *Ubx* domain of the Bithorax complex, *Cold Spr. Harbor Symp. Quant. Biol.* 50[1985]:181–94.)

abdomen. Remove the effects of BX-C, and the underlying genes for wing and leg production are free to operate in all these segments, turning them·into copies of the middle thoracic segment.

Such a dramatic effect is too much for the developing larva, which actually dies before it hatches from the egg. But before it dies, one can see on its surface the shape and pattern of the larval segments, which prefigure what the adult segments will become. It is thus possible to tell that if it were somehow to survive to adulthood, it would have an enormously extended thorax with twenty wings and twenty-two legs, and it would have no abdomen at all. The lower part of figure 8.2 shows what such an adult might look like.

Mutations of this type are called *homeotic* mutations. They were given that name at the end of the last century by the pioneer geneticist William Bateson (who also coined the term *genetics* itself). The term comes from the Greek, meaning "to make alike." The deletion of BX-C produces a classic homeotic mutation. It removes one level of genetic control from the development of a series of segments. Structures appear in inappropriate places, but they are not totally new. Homeotic mutations produce *mouches scramblées*.*

The Antennapedia complex is very similar in its genetic organization to Bithorax, and indeed they even occupy similar-looking parts of the giant chromosomes. It is now quite certain that the two complexes had a common ancestor in the remote past. But after all this time the functions of ANT-C and BX-C have diverged, for ANT-C controls a level of development one step further back. It specifies the developmental pathways that BX-C modifies.

Delete ANT-C, and you would end up with a fly (if it survived) in which the middle thoracic segment looked like the one just in front of it. It would have lost its wings, and the

---

* Franglais for scrambled flies. Once, in a pretentious but seedy hotel in the north of England, I found *oeufs scramblés* on the breakfast menu. I have been itching to drag that story in somewhere ever since.

middle legs would look like forelegs. Further, the head would be distorted, for the segments of the head would resemble those in front of them.

This shows that ANT-C is responsible for the properties of the middle thoracic segment, and to a lesser extent the properties of the segments in front of it. ANT-C is also active in the abdominal segments, and it is this activity that is revealed when BX-C is removed.

When the level of control imposed by ANT-C is removed in turn, this reveals the effects of yet another set of genes. These are the ones that shape the front part of the fly, and in the absence of ANT-C, their influence spreads farther back than it should.

So the development of the fly is like a medieval palimpsest, in which each text is written over an earlier, partially effaced one. Somewhere down at the bottom of all these genetic messages is the most primitive set of all, which specifies the basic ancestral segment pattern. Legs, wings, antennae, eyes, and other structures are all add-ons to this basic simple segment, repeated copies of which made up the bodies of the worms of the fly's remote ancestry.

The supergenes BX-C and ANT-C and the other related genes found elsewhere in the fly's genome form a developmental toolbox that is even more sophisticated than the toolbox that produced the glycolytic enzymes. And, like the glycolytic enzyme genes, the genes in this toolbox are related through duplication and shuffling of parts.

Dramatic as these homeotic mutants are, they are not as spectacular as some others in which you can actually see the effect in adult flies. One of these, the mutant Antennapedia for which the ANT-C complex was named, has legs instead of antennae growing out of its head.

We now have a fairly good idea of how this grotesquerie can happen. It occurs because of the overproduction of an ANT-C substance.

The critical study was carried out by Richard Garber and

his group at the University of Washington in Seattle. They reasoned that if deletion of ANT-C removes control, then turning-on of ANT-C in parts of the fly where it is usually not expressed should exert this control in inappropriate places. To investigate this possibility, they examined a particular *Antennapedia* mutant.

It was already known from studies of the giant chromosomes that this mutation was the result of a small inversion of a piece of the chromosome that included part of the ANT-C complex. They found that this inversion had the effect of joining the *Antennapedia* gene onto another gene (of unknown function) that was normally expressed in the head of the fly. As a result, the gene for making legs was expressed inappropriately in one of the head segments.

Stephan Schneuwly, Walter Gehring, and their collaborators at the University of Basel went a step further. They attached ANT-C to a regulator for a different gene in the fly, a gene that is only expressed if the temperature is raised briefly during development. Flies carrying this genetic construct raised at normal temperatures developed normally. But when such flies were heat-shocked as developing eggs, many of them had legs instead of antennae. The ANT-C control was indeed being exerted in the wrong place.

Many other homeotic mutants in Drosophila are known, some with quite startling effects. In *proboscipedia,* another mutant within the same complex, legs grow where mouth parts should be. A mutation on another chromosome, *engrailed,* makes the front part of the wing into a mirror image of the hind part.

A change in the environment during an insect's development can often mimic these genetic effects. The effect produced by amputation of the antenna quoted at the beginning of the chapter, though it occurs in a different insect, is similar to that produced by *Antennapedia.* It is possible to use such an environmental shock to mimic the effects of other homeotic mutants, and we will see some vivid examples of this later.

All this and more can happen in the insects. Do similar things happen in vertebrates like ourselves, or have we diverged too far from our common ancestor with the insects? The answer seems to be that we have indeed taken a very different path. In organisms like ourselves, changes in the environment during development can indeed produce dramatic alterations, as the recent dreadful experience with thalidomide demonstrates. But these effects cannot be mimicked by mutation in a single gene. There are no mutations in the human population that regularly produce deformities like those seen in thalidomide babies. The environment, aided by some more general properties of the particular genotype, must act as a trigger.

The environmental trigger in the vertebrates can be surprisingly mild. It is possible, for example, to treat a small part of the skin of a developing tadpole with a weak acid or alkaline solution and cause it to grow an extra head at the treated point. But, in contrast to the insects, there are no "extra head" mutations in the frog that produce the same effect as this environmental shock.

Indeed, in vertebrates very few clear-cut homeotic mutants have been turned up. Perhaps the best example found so far is a single gene mutation in the mouse that results in an extra pair of ribs. Close examination of the mutant mice shows that a cervical vertebra has been transformed into a thoracic vertebra, complete with its pair of ribs. This mutation is mild in its effects compared with the genetic upheaval that can be caused by mutations of the Bithorax complex.

THE BITHORAX GENE REVEALED

Most of what we have learned about the genetics of BX-C is the result of decades of patient work by Edward B. Lewis at CalTech. Lewis has unraveled an extremely complex story.

The bottom line is that BX-C is a giant gene with many parts that interact strongly with each other, usually predictably but sometimes in unexpected ways. Despite these complexities, by and large the gene *maps to the organism*. The first half of BX-C affects the thorax, and the second half affects the abdomen.

The Bithorax complex has now been cloned by a most ingenious technique called *walking* down the chromosome. Welcome Bender and David Hogness at Stanford University were the inventors of this amazing technique. Starting with a cloned piece of DNA that they knew was close to BX-C, they then fished in the genome for pieces that overlapped the clone. They used clones of these pieces in turn to fish for pieces that overlapped *them*, and so on down the chromosome. When they reached BX-C, they found that it was an enormous piece of DNA some three hundred thousand bases long.

Once they had cloned and mapped this region, they discovered that many of the mutants Lewis had spent years collecting are due to deletions or inversions of the DNA. But many others are due to the insertion of jumping genes. For example, classes of spontaneous mutants called *bithorax* and *bithoraxoid* are almost all due to the insertion, at various points in the BX-C, of a mobile element called *gypsy*. A few other elements have been found to be responsible for some of these mutants, but gypsy is by far the most common. It is one of a family of elements, with a strong resemblance to a group of mammalian viruses called retroviruses, that are capable of inserting copies of themselves into various places in the genome. Gypsy has also been found to be responsible for many other mutations in developmentally important genes of the fly. But not in all.

Mutations at the *white* locus, for example, produce flies with white or pale eyes rather than the normal red. More than half of the spontaneous mutations that have been investigated at this locus turn out to be due to the insertion of different mobile elements with names like *copia*, *P*, and *412*. At the *notch* locus, which affects the wings and the eyes, two other elements related to copia are responsible for many of the mutants. Gypsy has not been found so far at either of these loci. It seems that different mobile elements have preferred settling places in the genome.

We see here again, as we have so often, a fascinating interaction between the genes of the organism and the mutational agents that shape them. It is apparent that in Drosophila, as well as in bacteria, different mobile elements have been shaped by a variety of selective forces. Partly as a result, they tend to settle into different parts of the genome. Once there, they cause specific types of mutations.

HOMEOTIC GENES AND EVOLUTION

Fascinating as all this is, and indeed it is at the very cutting edge of developmental biology at the moment, what does it have to do with evolution? Do I really imagine that the pitiful gallery of grotesques produced in the BX-C by deletions and the insertion of *gypsy* elements have any evolutionary importance at all? What could possibly be the advantage of a fly with extra legs or wings? Obviously, little or none. These mutant flies have great difficulty just surviving to adulthood, and in many cases cannot even manage that.

Further, the homeotic mutants are not producing anything *new* but are merely scrambling preexisting parts of the flies. Surely, if the organization of the genome somehow lends itself to evolutionary change, it should be capable of producing new structures, not simply rearranging the ones that the organism already possesses. In fact, however, most evolution proceeds through just such rearrangement, followed by gradual modification of the rearranged parts. While it is difficult to imagine a fly with legs on its head surviving at the present time, it might be that in the past such a *possibility* for rearrangement might have had a great advantage.

The remarkable one-to-one mapping of parts of the BX-C and ANT-C to the structure of the insect itself means that we can think of the insect as a series of genes made visible. Changes in the sequence of the genes will often produce corresponding changes in the sequence of structures that make up the insect. It is as if the insect's entire body is the expression of a giant chromosome made visible to the naked eye. Of course, the correspondence is not perfect, and the genes we have talked about are only the top layer of a web of regulatory genes. But the correspondence is there, and it is far more pronounced than it is in the vertebrates.

The appendages of present-day insects—the wings and wing cases, eyes, legs, antennae, mouthparts, balancing organs—are all modified from flaps of integument and other simple structures possessed by their primitive wormlike ancestor. Their divergence into different functions had to start somewhere, and it began with the process of segment duplication.

We saw in chapters 5 and 6 that the duplication of genes is a powerful generator of new evolutionary possibilities, since the duplicates are now free to evolve in different directions. In the insects, the duplication of body segments is equally pregnant with possibility.

Mutants affecting segment duplication in Drosophila are some of the most dramatic developmental mutants yet found. There are about ten of them scattered around the chromosomes. Most of these have the effect of producing larvae in which alternating segments are missing, a change which proves to be fatal early in development.

One of these mutants is found in the Antennapedia complex. It is a gene called *fushi tarazu*, which is Japanese for "not enough segments."*

It is obvious from the appearance of the mutant larvae that the *fushi tarazu* gene is being expressed in alternating segments, because when the gene is destroyed the segments disappear. But what is it doing in the segments that are left? Is it expressed and masked by some other gene that permits the segments to develop, or is it not expressed at all? If the former, then if the gene is turned deliberately on throughout the fly, nothing should happen. If the latter, then if the same thing is done, something interesting might take place.

Gary Struhl, working at Harvard, succeeded in turning on the *fushi tarazu* gene throughout the fly by attaching it to the same heat-stimulated gene used by the Basel group to investigate ANT-C. Something interesting did indeed happen. When he subjected the eggs of these flies to a heat shock during development, the *other* set of segments did not develop. This showed that the *fushi tarazu* gene must be turned *on* to let one set of segments develop, but it must be turned *off* if the alternating set is to develop.

A remarkable picture is emerging. In the circumscribed world of adjacent segments, the *fushi tarazu* gene acts in much the same way that ANT-C acts on the whole fly. You will remember that ANT-C has to be turned on in the thorax to make the wings and legs of the thoracic segments but that it has to be turned off in the head segments if these are to develop normally.

There are some other similarities as well. Suppose it were possible to switch the roles of the *fushi tarazu* gene so that it is turned on and off in all the wrong segments. Would the developing larva vanish completely like the Cheshire cat?

Not quite. It would leave the equivalent of the Cheshire cat's smile behind. The segments that are turned on and the

---

* This is one of the few Japanese phrases familiar to all geneticists. I cannot think of much use for it in everyday life, except to complain of a niggardly serving of rolled sushi in a Japanese restaurant.

segments that are turned off are slightly out of register. This is just like ANT-C, for the region where ANT-C is expressed overlaps the region where the head genes are expressed.

The mode of action of the *fushi tarazu* gene is hauntingly similar to that of ANT-C and BX-C, and indeed they probably had a common ancestor. But its message is written on another, deeper layer of the developmental palimpsest, and we are nowhere near the bottom layer yet.

ANT-C and BX-C give us a glimpse of the past in which the ancestor of Drosophila consisted of a series of similar, fairly undifferentiated segments like many of today's worms. But *fushi tarazu* and the other so-called *pair-rule* genes provide a glimpse of an even remoter time, when the ancestor had fewer segments than Drosophila does now.

Imagine that this wormlike ancestor had half as many segments as the fly does now. Duplication of a gene ancestral to *fushi tarazu* that controlled segment development might result in a sudden doubling of the number of segments. This mutational event would have been even more dramatic in its consequences than the sudden doubling of the chromosomes that occurred in our fish ancestors because it would produce a large alteration in the worm's appearance. Were such a doubling to occur in a fly now, the consequences would certainly be fatal. But had it happened when the ancestor of the fly was a relatively undifferentiated wormlike creature leading a simple burrowing existence, the creature might have survived. And the alteration might even have been advantageous. The pair-rule genes like *fushi tarazu* may reflect the traces of a *real* hopeful monster event that took place in the distant past.

This doubling event, if it did in fact occur, must have taken place back in the "Model T" period of evolution. Similar but smaller events can and do take place now. Individual segments have often been gained or lost in the course of insect evolution. Such relatively small changes, even if they occur suddenly, would presumably have much less impact on fitness than a sudden doubling of segment number.

The possibilities for segment evolution in the insects are very great, but they are not unlimited. Judging by what geneticists have turned up, the developmental genes of Drosophila seem unable now to produce mutants that are, by any stretch of the imagination, an improvement on the fly's body plan. Perhaps the last hopeful monster event in the ancestry of Drosophila was when its four-winged ancestors lost two of their wings. But this apparent lack of present-day hopeful monsters may simply reflect our lack of imagination. It is of course unlikely that flies with legs instead of antennae would ever survive, but the Drosophila literature abounds with the discovery of less drastic homeotic mutants. Some of them, somewhere, at some time, might be advantageous.

I suggested in chapter 6 that the glycolytic enzyme toolbox is probably closed by now. The Drosophila developmental gene toolbox is still open a crack, but the types of successful mutants it can make are limited. This is because the development of insects is still constrained by their compartmentalization. Many evolutionary possibilities have as a result been denied to them. Their express train is steaming down a track made plain by signals very different from those that mark off our own.

Even though they are the most successful of animals, the insects will have to break down the walls of their compartments if they are ever to take a new evolutionary direction. In the meantime, as we saw with the butterflies and as is daily apparent in the world around us, they have become very good indeed at what they do. And that appears to include evolving in certain well-defined and thoroughly practiced directions, and avoiding evolution in others.

9

The Increasing Sophistication
of Evolution

. . .What had that flower to do with being white,
  The wayside blue and innocent heal-all?
What brought the kindred spider to that height,
  Then steered the white moth thither in the night?
  What but design of darkness to appall?—
If design govern in a thing so small.
                                    —ROBERT FROST, "Design"

A "CREATIONIST" EXPERIMENT

I promised at the beginning of the book that our examination
of some of the many ways in which evolution has been facilitated would have an intellectually pleasing result. This
would be to make evolution appear both more understandable and more *likely*.

We have now gone through enough cases of evolutionary
facilitation to be (I hope) fairly comfortable with the idea and
to glimpse some of the ways in which the likelihood of certain
kinds of evolution has increased. Let me go back now to a

very unsophisticated kind of evolution, the sort that might have occurred during the earliest stages of life on the planet, and show how even that is more likely than it first appears.

Just how unlikely is evolution? It depends on what kind of evolution you mean, and on whom you talk to. To a creationist, the process seems impossibly unlikely. This is because, in order to construct a straw man, the creationist will pick what appear to be the most unlikely features of evolution—the appearance of complex organs, behavior patterns, and molecules, for example. And to make the argument stronger, the creationist will assume that evolution works in the clumsiest and most unsophisticated fashion imaginable. Let us follow out a typical creationist argument.

You will recall my mentioning a few chapters back that if you simply string amino acids together in the laboratory to make a *random* protein, you will get a contorted, ugly molecule with few if any of the elegant helices and pleated sheets that characterize the proteins making up our body. Such proteins are unlikely to have much if any detectable function. The overwhelming majority of random proteins will fall into this category. I can illustrate this with a small variation on the familiar "monkeys typing Shakespeare" scenario.

Imagine a vast army of monkeys trained to operate protein synthesizing machines rather than typewriters or word processors. Suppose that there are enough of these monkeys to fill up the entire universe, and that each monkey and its machine takes up 10 cubic meters of space. Suppose further that each of these monkeys can produce a protein made up of a chain of 347 amino acids in an hour. The machines are built in such a way that there is an equal probability of any of the 20 amino acids specified by the genetic code being inserted at a given point in the protein. How long will it take before a monkey produces a protein with exactly the same amino acid sequence as the yeast alcohol dehydrogenase?

It might happen right away. Things considerably less strange but nonetheless unlikely do happen. A young woman once told me that she had won the triple perfecta at the very first race meet she had ever attended. She bought a vacation home in Portugal with the proceeds and very sensibly never went to the races again.

But the odds against picking the winners in all three of those races were considerably shorter than those facing our monkey army, even though the idea is the same. It is quite possible that one of the legion of monkeys will produce the "right" protein on its very first try, but the chances are very small.

If there are 20 possibilities for the first position in the protein, 20 for the second, and so on, then the number of all the possible proteins 347 amino acids long is given by 20 × 20 × 20 . . . 347 times, or 20^{347}. So, the chance of making the right one the first time is only one in 20^{347} or one in 10^{444}, an exceedingly, *exceedingly* small number.

Perhaps if we get enough monkeys together, this will help. How many monkeys can we fit into the universe, each in a comfortable cubicle of 10 cubic meters?

In making our calculation we will ignore the fact that there is nowhere near enough matter in the universe to make a sufficient number of monkeys and machines so that one of each can be put in every 10 cubic meters.* We will proceed with the calculations as if there were.

If the universe is a closed one and did indeed start with a big bang, then it is possible to estimate its current "circumference" at about 125 billion light years. This is how far one would have to travel to get back to the place one started. So, we can think of the universe as a sphere with a radius given by this circumference divided by two pi, and with ourselves in the middle.

Such a sphere will have a volume of about 10^{33} cubic light years. A cubic light year is very big, and it is possible to fit

* And if there were, the universe would be a lot smaller than it is. But such quibbles never trouble people who make monkeys-typing-Shakespeare calculations.

about 10^{38} little cubicles in it, each with a volume of 10 cubic meters. This gives us a grand total of $10^{33} \times 10^{38}$, or 10^{71} cubicles, each with a monkey and a machine.

What is the chance that somewhere in this immense agglomeration of monkeys one of them will get it right the first time? Only, alas, one in $10^{444} \div 10^{71}$, or 1 in 10^{373}. This number is certainly a lot smaller, but in practical terms we have hardly made a dent in the odds.

Now, it can be shown quite simply that the *average* number of tries a monkey will have to make to get the right sequence is 10^{444}. This is because, while it might be lucky and get it right the first time, it might be extremely unlucky and keep making the wrong proteins over and over. There is, of course, no prohibition against its making the same sequence more than once. An unlucky monkey might labor two or three or fifty times as long as the average to make the right sequence. Even if the universe full of monkeys managed to do it in the average amount of time, it would still take them 10^{369} years (since there are about 10^4 hours in a year).

To put things in perspective, it is calculated that the universe is somewhere around 2×10^{10} years old. It will have come to an end, one way or another, long before the monkeys finish their thankless task.

Now, this is the essence of one of the most powerful of the creationist arguments that have been leveled against the proponents of evolution. There are such an immense number of possibilities for even the simplest biological structure that it stretches likelihood to suppose that any of them could have happened, even given all the time that the universe has been in existence.

Of course, as many evolutionists have pointed out, this argument ignores the process of evolution itself. If it is to an organism's advantage to make or eat alcohol, then *any* protein that the organism happens to have that acts on alcohol will be selected for. Any improvements on this molecule will also rapidly be selected for, because the genetic information encoding this protein and its subsequent improvements will be passed from generation to generation. Natural selection is a cumulative process, carrying with it the improvements that have gone before.

This is the standard answer to the creationist argument, and it makes the point that evolution is a good deal more sophisticated than a bunch of monkeys pounding typewriters or cranking out proteins. But there is another point to be made, one perhaps not so widely appreciated.

If you think it is difficult for the monkeys to produce the alcohol dehydrogenase, consider how unimaginably more difficult it would be for them to produce *Hamlet* by typing at random. The play is much longer than 347 letters, and there are many more than 20 symbols to choose from (upper and lower case letters, punctuation, spacing, and so on).

But if we set an unimaginably large number of universes full of monkeys to work, sooner or later one of them will succeed. And along the way various monkeys will type lots of other things. Many of the unauthorized versions of *Hamlet* that were scribbled down in the pit of the Globe and hastily printed by piratical publishers will appear. So might *Death of a Salesman*. And so will many of the plays that Shakespeare toyed with the idea of writing but never got around to. *Son of Hamlet*, perhaps.

The point is that for anyone with the patience to wade through universes full of garbage, the monkeys' output would be a treasure trove of great literature. Some of it greater than *Hamlet* itself.

Now, it turns out that a number of scientists have in fact carried out the creationist experiment, though on a far smaller scale than the production of a play or even of a whole protein molecule. And they have shown that there are remarkable biological treasures concealed in the universe of random molecules.

Several groups have done or are doing experiments along these lines, but perhaps the closest in conception to the monkeys and the typewriters was carried out by Marshall Horwitz and Lawrence Loeb at the University of Washington in Seattle. They set out to find whether they could select for a functioning bit of DNA from a collection of essentially random molecules of a given length that had been inserted in the place of an original sequence.

The original DNA that they started with is a regulatory element called a *promoter sequence*. This particular promoter is positioned a little way "upstream" from a gene that confers resistance to the antibiotic tetracycline on the *Escherichia coli* bacteria carrying it. If the promoter sequence is disturbed or destroyed, messenger RNA can no longer be made from the gene, and the bacteria die in the presence of even low levels of tetracycline. But the more efficiently the promoter sequence works, the more messenger RNA and protein are made and the more resistant the bacteria become.

Horwitz and Loeb removed a nineteen-base sequence from the promoter region, essentially destroying its function. They then made an enormous number of nineteen-base sequences of DNA at random, using a machine that can synthesize short stretches of DNA. Ordinarily, this machine is programmed by its operator to attach bases to the growing chain in a given sequence, but this programming ability can be short-circuited by giving it a mixture of all four bases to use at each step in the synthesis. Because even a tiny amount of DNA contains many molecules, Horwitz and Loeb could easily generate most of the 4^{19} (roughly 10^{11}, or one hundred billion) different possible molecules, even though each one might have been represented only one or a few times in their sample.

They gave their random molecules two highly specific "sticky" ends, one at the beginning and a different one at the end. These matched sticky ends in the DNA from which the promoter region had been deleted. Thus, the DNA could not rejoin unless one of the 10^{11} different little random pieces had been inserted. The little pieces could only be inserted in one orientation to heal the gap, and only one little piece could be inserted into each molecule carrying the deletion. (For an illustration of this process, see figure 9.1.)

The next and most exciting step was to see whether any of the *E. coli* carrying the little random pieces were resistant to tetracycline. Horwitz and Loeb found many different such mutants, and on sequencing them they discovered that all had a resemblance—though sometimes a very vague resemblance—to the original stretch of DNA that they had deleted. All were good promoters, and some were actually better than the original.

It even turned out to be possible to make promoters starting with random families of DNA constructed by giving the machine mixtures of just three or even just two of the four bases. Some of these promoters showed no resemblance at all to the original sequence that had been deleted, but even so some of them were better than the original promoter. This was in spite of the fact that the number of possible nineteen-base sequences in the random families was drastically curtailed by using fewer than four bases as building blocks. In the case of the two-base promoters, the number of possible sequences is only 2^{19}, or about half a million.*

This remarkable experiment opens up exciting possibilities. By using selection to sort out random pieces of DNA, it should be possible for experimenters to make startling evolutionary leaps to genes with properties that have simply not been achieved in the living world.

At the same time, the experiment emphasizes some important facts about evolution. The creationist experiment, when it is done properly, demonstrates that even the clumsiest kind of evolution is not so unlikely after all. There really are many different ways to solve an evolutionary problem in that imaginary universe of different genes. Some of them may be better, indeed far better, than anything evolution on

* Because of the way the experiment was arranged, nowhere near this many possible combinations were actually tested. There are indeed many treasures concealed in this collection of random sequences of DNA.

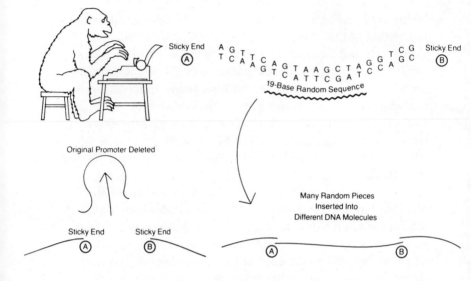

Original Promoter Sequence:

T C A T G T T T G A C A G C T T A T C

Some of the Random Sequences That Work:

G G C G A T G C T [T] T C C A C [T] A [G] A [T] T A G
T T G G G C G C G C G T C G [G] C [T T G]
             G G G C G [G T] C [T] C C [C] G [G] T C G T [T]

(Parts That Are the Same As the Original Promoter Are Boxed)

Figure 9.1 A diagram showing how Horwitz and Loeb carried out their "creationist" experiment. They produced many different 19-base random sequences, each supplied with sticky ends A and B. These could then be inserted in place of the original 19-base piece of promoter, which had been deleted, leaving sticky ends that matched A and B. A few of the random sequences that worked are shown at the bottom of the figure. These have been lined up so that the parts that look most like promoters in general are matched. You can see that these sequences do not greatly resemble the original promoter, though they worked at least as well as the original. (One of the new promoter sequences does not have 19 bases, but this is a consequence of the way the experiment was done.) (Adapted from part of figure 2 of M. S. Z. Horwitz and L. Loeb, Promoters selected from random DNA sequences, *Proc. Natl. Acad. Sci. U.S.* 83 [1986]: 7405–9.)

this planet has been able to come up with. But they may not be reachable now, because the interim solution arrived at by present-day organisms is quite serviceable. Getting to the better solution might involve dismantling the present-day gene and replacing it with something utterly different, as Horwitz and Loeb did.

Many of the mutational and developmental processes we have talked about in this book can be thought of as the evolutionary equivalent of monkeys at their typewriters or protein-synthesizing machines. But in the systems that we see about us now, the monkeys are constrained and prevented from producing completely random nonsense.

The human immune system manufactures enormous numbers of antibodies that never come into play because they do not meet the appropriate antigen. But it does not produce random pieces of protein. In the case of the immune system, it is as if the monkeys were working at computers instead of at typewriters, endlessly rearranging different combinations of blocks of text from *Hamlet*. The play would emerge far sooner from such an operation than if the monkeys were manipulating individual letters or even words. But we would never get *Son of Hamlet*.

Making an enormous amount of garbage to produce a few usable genes is obviously a very inefficient way to proceed. But at the outset of the evolution of life it may have been the only way. In the process of becoming more efficient at evolving, the sophisticated genomes of the present have perhaps lost some of the daring adventurousness possible during the early days, when genomes were simpler and closer to the monkeys and their typewriters.

You will recall our trip to the Precambrian, and the simple organisms we found there. One of the reasons for the exuberant explosion of different life-forms that followed shortly thereafter during the Cambrian might have been because the constraints on those organisms were so much less than on organisms living at the present time. Even a clumsy organism moving into a new niche in those days had some chance of surviving.

But where did all this variation come from? If we could only get hold of the genes of those Precambrian creatures, we might find that they were much closer to the monkeys and the typewriters than our own sophisticated genomes. Those ancient organisms might even have possessed mechanisms for producing and juxtaposing random DNA sequences, because the advantage of producing totally new genes in order to invade new and utterly unoccupied niches would have been so great.

We may find the traces of these ancient mechanisms in our own genomes, once we know where to look. But they are unlikely to be more than traces, for our genomes are now capable of changing in far less effortful ways. They have also erected strong barriers to undesirable change. We have come a very long way from the monkeys and their typewriters, and in the process we have both lost and gained.

THE SNAP-OUT GENES OF SNAPDRAGONS

The growing sophistication of the evolutionary process is most vividly displayed by the prevalence of what I have termed evolutionary toolboxes. These toolboxes themselves vary in complexity from simple ones of fairly recent origin to much more elaborate ones that are intimately connected with the survival of the species.

In shaping domesticated animals and plants over the centuries, our own ancestors stumbled on a great many evolutionary toolboxes. Plant breeders were particularly successful in finding and using them. For example, it took the efforts of many generations of pre-Columbian Indians to select the mottled and variegated kernels of Indian corn. These highly variable strains, you will remember, later provided Barbara McClintock with some of the clues she needed to unravel the story of jumping genes.

One very simple toolbox that has recently been investigated at the molecular level produces variegated flowers in the snapdragon.

The snapdragon *Antirrhinum majus* has a rather pretty flower consisting of a tubelike corolla with lobes radiating from the tip. Its common name in English suggests a similarity to the face of a dragon. This can be perceived with some effort of the imagination. It received its scientific name, however, from a supposed resemblance to a calf's nose, a resemblance that I for one cannot detect.

Antirrhinum has been cultivated for many human generations. Like other domesticated flowering plants, its flowers have been selected during that time for increased size and for unusual colors and patterns. The most common variety has flowers of a plain dark red, with some yellow at the base. But many other types have appeared spontaneously and been selected for.

One of these is a pale flower covered fairly uniformly with small red spots. It turns out that this pale background color is the result of a mutation that destroys the ability of the plant to make an enzyme involved in the production of the red pigment. The red spots are in turn caused by repeated genetic changes back to the ability to make the enzyme, restoring the mutant cell's pigment-producing ability. (See figure 9.2 for two types of snapdragons.)

These back mutations take place while the cells that are to make up the flower are multiplying. During the last stages of flower growth there is an enormous increase in the size of the cells, so the effect is as if you made red dots all over an uninflated balloon with a pen, then blew the balloon up. The dots would all still be there, but each would now have become a patch.

Somatic mutations such as these, you will remember, take

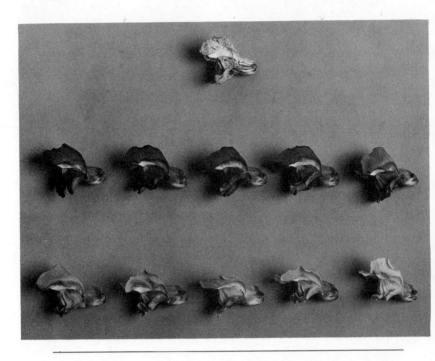

Figure 9.2 The spotted flower of *Antirrhinum majus* on the top shows the effects of repeated removal of the Tam3 element, resulting in spots of red pigment on a pale background. The flowers in the two rows below show the range of colors that can be obtained after the element has jumped out of the chromosome, leaving various-sized bits of itself behind. These flowers, which now breed true, range from deep red to pale pink. (From figure 1 of E. S. Coen, R. Carpenter, and C. Martin, Transposable elements generate novel spatial patterns of gene expression in *Antirrhinum majus*, *Cell* 47[1986]:285–96.) Copyright © 1986 by Cell Press.

place in the body of the plant and not in the cells that will be passed on to the next generation. Since there are hundreds of spots on each flower, this means that the mutation rate to restored pigment is astonishingly high. Further, the rate is strongly dependent on the environment. Plants grown outside a greenhouse have many more spots per flower than those grown inside it, where the temperature is higher.

Given what we have learned so far, it does not take Sherlockian powers of deduction to suspect the work of jumping genes. Indeed, once the molecular tools became available, workers at the John Innes Horticultural Institute in England quickly discovered that this high somatic mutation rate was due to an event that occurred again and again in the flower cells. This was the removal of a transposable element. The element, though slightly different from those found in bacteria and fruit flies, has many points of similarity. It is called Tam3 (Tam is short for *T*ransposon *A*ntirrhinum *m*ajus).

The effect of mobile elements on the genes into which they are inserted is often difficult to understand. The gypsy elements we talked about in the last chapter, for example, have complex and ill-understood effects on the BX-C. But the mode of action of Tam3 is much simpler. *Antirrhinum* normally has several copies of Tam3 inserted at different places in its genome. At some point in the past a new copy was inserted into a particular part of the genome of one plant, giving rise to a pale-flowered mutant. This was because this new copy of Tam3 nestled a little way in front of the gene coding for the enzyme responsible for making the red pigment. It disrupted a promoter region, and as a result the gene could no longer work properly.

This change might be expected to be relatively stable, like most mutations once they have occurred. But instead, the element tends to jump out of its place at a very high rate. Each jump results in the restoration of the ability to make red pigment, and each red spot on the flower represents a separate jumping-out event.

This jumping is not confined to the flowers, but can also take place in other parts of the plant. The red pigment is only made in the flowers, however, so when a jump occurs in the cells of other tissues, it leaves no visible trace.

Sometimes the jump occurs in one of the cells that is slated to become a sex cell and to send its genes on to the next generation. At other times, Tam3 does not jump out but rather rearranges the genetic information in which it is embedded, causing changes in the numbers of spots in sub-sequent generations. If Tam3 jumps out cleanly, all the cells of all the flowers of the new plant will show the effects of having lost the element. They may, for example, all revert to the original red color. Or, if the element has left little bits of itself behind, this may lower the amount of red pigment that is made. This is because the little bits of the element that are left behind damage but do not destroy the promoter. The result of such an incomplete removal of Tam3 will be a color that lies somewhere between red and the ivory background of the previous generation. All the flowers of subsequent generations will retain this new color, and none will be spotted.

Most of the time, Tam3 leaps out of its place intact, but imperfect removal and genetic rearrangement occur fairly commonly. The result is a wave of genetic changes that take place at frequencies much higher than could ever be achieved by ordinary mutational agents.

The changes produced by this copy of Tam3 are confined to those parts of the genome where it originally settled. If they were not so confined, they might inflict high levels of genetic damage of the kind caused in *E. coli* by bacterophage Mu. But because they are confined, they produce a fairly narrow range of quite harmless and often attractive alterations in the appearance of the flowers. The original Tam3 insertion must have occurred at some point in the recent past. And because it produced ivory flowers with red spots, horticulturists immediately picked out the plants carrying it. Repeated excision events subsequently produced an explosion of interesting varieties of flower that were carefully preserved by enthusiastic gardeners.

But the gardeners did not cause the original insertion event; nor did they produce the Tam3 element itself. Both Tam3 and the genome it inhabits have been shaped by selection that greatly predates the evolution of gardeners. The properties of this element are the result of a long evolutionary history of interactions between the element and the plant it inhabits. We do not yet know the real role of Tam3 and its importance in evolution, but we are starting to have our suspicions.

These suspicions have to do with parts of the genome that we do not yet fully understand. It might be imagined by this time that geneticists have unraveled essentially all the mysteries of the genes. And we have learned a huge amount from our understanding of DNA, of the role of RNA, and most recently of the ways in which gene products bind directly to the DNA to regulate other genes. In fact, however, each gene, no matter how well studied, continues to hold its secrets. And there are huge categories of genes that we do not understand at all.

One of these categories governs characters familiar to all of us. These are genes that produce continuous rather than discrete variations in the appearance of organisms. Genes that control height in humans are one excellent example. There are many of them, and they interact so strongly and unpredictably with each other and with the person's environment as he or she is growing up that it is not possible at a child's birth to say much more than that it is likely to be tall or short. This is *multigenic variation*. Because these genes individually have small effects, often masked by those of other genes or the environment, they have been difficult to clone and study in detail.

Recall that if Tam3 is not removed from the chromosome perfectly, but leaves little bits of itself behind, it can produce genetically stable flower color variants. All these changes, producing a range of intensities of color, are due to alterations in the promoter of a particular gene. But if Tam3 and other mobile elements were similarly active elsewhere in the genome, they might be responsible for producing a good deal of the multigenic variation that makes the inheritance of so many genetic traits difficult to understand.

In short, if these elements are really operating all over the genome and can affect several different genes that contribute to a character, they may be rapid and powerful producers of genetic variability.

There are two relevant points that emerge from this story. First, horticulturists have been able to pick out a new genetic configuration that has made subsequent selection of interesting flower types easier. This configuration is an evolutionary toolbox that can produce snapdragon color and pattern variation. Nature has often selected for such toolboxes, and the result has been to make certain evolutionary pathways easier.

Second, this toolbox is very simple and crude. These mobile elements leave little bits of themselves behind and in the process generate new genetic variation in the same way that a child leaves muddy footprints on a clean kitchen floor. Occasionally, an aesthetically pleasing pattern appears on the floor, but not often.

The Tam3 toolbox has progressed some distance from the monkeys with the typewriters because the changes it makes are not completely random. It still retains many monkey-typewriter attributes, however, because the changes it makes still have a large random component.

This is not the case with many more specialized toolboxes.

THE CUNNING TRYPANOSOME

The sudden appearance of the AIDS virus in the human population of Africa, probably no more than two decades ago, illustrates the fact that the tropics are a repository of rare diseases that can suddenly spread without warning. Before they spread, these diseases might have been confined to animals in remote and unpopulated areas, as was apparently the case with AIDS. Or they might have affected only isolated tribes of humans with few or no links to the rest of the world.

This latter situation appears to have been the case with

African sleeping sickness. An insidious disease, sleeping sickness may take two or three years to develop. It begins with an almost trivial set of symptoms, including headache and lassitude. As the disease develops, rashes, fever, and edema follow each other, and the patient progressively loses interest in his or her surroundings. Until recently the final coma was inevitably followed by death. Even now, if the disease is allowed to progress too far before treatment, severe brain damage may result.

We do not know the original distribution of the disease, although it was probably not extensive. But during the eighteenth and nineteenth centuries, rapacious slavers, Christian and Muslim, drove into the African interior from the west and east coasts. This upset a precarious balance. The disease began to spread to previously unaffected tribes.

This spread was accelerated by the inroads of the missionaries and explorers, who hired hundreds of the local inhabitants to haul their massive Victorian paraphernalia from one side of the continent to the other, bringing the disease with them. Huge forced evacuations were carried out by colonial administrators in an attempt to control the subsequent outbreaks. The largest of these was the evacuation of one hundred thousand people from the shores of Lake Victoria in 1907. While these shifts appeared to solve the immediate problem, they were also instrumental in disseminating the disease even more widely.

Cattle were introduced into the interior. Whole groups of tribes rapidly became dependent on them. Just as rapidly, this new form of wealth fell victim to a disease called *nagana*, closely related to the human form of sleeping sickness.

The herdsmen of some South African tribes quickly noticed the relationship of nagana to a fly called *tsetse*. These flies, of the genus *Glossina*, are distinguished by a rigid projecting proboscis, which they use to slash the skin and drink the blood of their hosts. Nagana tended to be worst in the marshy areas where the flies abounded and disappeared when the fly season was over. This connection was confirmed by the work of David Bruce, an Australian microbiologist, who showed at the end of the last century that the disease was indeed transmitted by the flies but that it was actually caused by a tiny protozoan parasite of both the fly and the mammalian host, called a *trypanosome*.

The trypanosome is an advanced single-celled organism, not a bacterium. It has properties very like those of the cells that make up our own bodies. This makes the disease particularly difficult to cure, for drugs that poison the trypanosome also tend to poison the host. It corkscrews its way rapidly through the blood, driven by a long flagellum that runs the length of its body. The name itself means "borer," and it can literally bore its way through tissue into the spinal fluid, where it multiplies and brings on the final stages of lassitude, neurological degeneration, and death.

David Bruce, Count Aldo Castellani, and David Nabarro are all now credited with the codiscovery of the slightly different trypanosome, spread by different species of tsetse, that causes sleeping sickness in man. The three of them wrangled for decades about who had actually made the discovery, a dispute that now seems as remote as the argument between Burton and Speke over who had discovered the sources of the Nile.

The disease and recent disturbances of the environment have together rendered large areas of equatorial Africa uninhabitable to man and his domestic animals. Remarkably, however, trypanosomes similar to those that cause sleeping sickness and nagana thrive in the blood of many large animals native to these areas—ungulates, members of the pig family, and even large reptiles. They sometimes cause fatal disease in these animals, but their effects are usually slight. This relative lack of effect on the indigenous animals illustrates rather vividly that man and his cattle are newcomers who have yet to coevolve along with the trypanosome to a mutual accommodation. But the potential for that coevolution is there, as an intensive study of the sleeping sickness and nagana trypanosomes is beginning to reveal.

Some of the other parasites we have talked about in this book can hide from the host's immune system. Malarial plasmodia spend most of their time within red blood cells. Schistosome worms coat their bodies with proteins from the blood of their hosts. But the trypanosome spurns these disguises and swims freely in the blood plasma. As a consequence it is continuously exposed to the attack of macrophages and the complement system. Because the trypanosome invaders can immediately be detected as nonself, the host will quickly make antibodies against them.

The result is that after about a week most of the trypanosomes in the blood are destroyed. The few survivors, however, are resistant and rapidly build up in numbers. Cells of this new population are resistant because the major glycoprotein on their surfaces has changed. Glycoproteins are proteins to which complex polymerized sugar molecules have become attached in the course of their synthesis. They form an important part of the outer surface of the trypanosome's cell. As a result of the switch, the host must now manufacture new antibodies to attack them.

This seesaw battle between the host's immune system and the continually altering parasite can go on for years as the infection runs its course. Over a hundred antigenically distinct types of parasite can arise, each from the one just before it.

While certain antigenic types tend to appear early in the infection and others later, the types do not appear in a precise and repeatable sequence during each infection. Such a predictable order might allow the host's immune system to evolve to meet the challenge. The unpredictable order of these changes is so effective in flummoxing the immune system that even the systems of the native hosts are defeated. These hosts are as powerless as man or cattle in eradicating the constantly evolving parasite. The accommodation that has evolved between the native hosts and their parasites appears to have come about primarily because these parasites are less likely to cause damage to their host's central nervous system than are those that infect humans and cattle.

Many of the sudden antigenic switches are accompanied by, and are probably the result of, distinct rearrangements of the trypanosome's genes. There are over one hundred different genes in the trypanosome's genome coding for different surface glycoproteins. But only one of them is turned on at any one time. If a new gene is moved to the appropriate place next to a regulatory region that allows it to be transcribed into RNA, it can be turned on and the old one turned off. The old gene, displaced from its favored location by the arrival of the new one, lapses into inactivity.

These shifts have been detected by the use of probes made from cloned antigen genes that are normally activated early in the infection. Mapping of the DNA in the region surrounding the gene that binds to the probe shows that the trypanosome's copy of the gene changes its position as the infection progresses. This alteration is passed on to all the trypanosome's progeny.

All these sequential alterations must somehow be temporary, in the same way that the construction of specific immunoglobulin genes in our somatic cells is only temporary. The trypanosome's genome must be "reset" in some fashion as it passes through the tsetse fly, so that the whole cycle can be repeated with variations in the next mammalian host. Nothing is known about this process, however.

Regardless of the details of how it works, this mechanism is superbly adapted to a rapid and repeated outflanking of the host's immune system. There is an absolute requirement that the structural integrity of the glycoprotein genes be retained, since the trypanosome's survival depends on them. But within these limits there will be very strong selection for diversity among this set of one hundred or so genes. The greater the diversity, the more work the host's immune system must do to manufacture new antibodies. The number of genes is great enough to allow a parasitic infection to persist for years, which gives plenty of time for the parasites to be spread to new hosts by the tsetse fly.

The switch from one gene to another is certainly a kind of mutation because it involves an alteration in the parasite's genes. But how restrictive a set of mutations! The changes occur in approximately, but not precisely, the same order in each succeeding infection. The whole system is as narrow, restrictive, and superbly adapted to the survival of the infective agent as is the system of somatic mutations that produces the diversity of antibodies in the host. Indeed, the progressive refinement of both the trypanosome's switch system and the host's antibody system may each have been spurred in the past by the continued evolution of the other. It is fascinating that both systems have so many points in common.

Mutational switching from one to another of a set of multiple genes or gene fragments appears to be so efficient a mechanism for generating adaptive diversity that it has appeared again and again in the course of evolution. Many other parasitic organisms, such as the bacterium responsible for gonorrhea, have evolved similar systems that increase their antigenic diversity. All involve the fairly regular shifting of genes from one part of the genome to another. All these schemes can only work because the genes are arranged in highly specific ways that facilitate the shift. Further, their evolution has been stimulated by the existence of the immune system with its very similar mechanisms for generating diversity. These collections of genes form a far more sophisticated type of evolutionary toolbox than we saw in the snapdragon.

TOOLBOXES AND THE EVOLUTION
OF ACQUIRED CHARACTERS

Arthur C. Clarke once said that any sufficiently advanced technology will appear to be magic to those who have never seen it. We are only beginning to be aware of evolutionary toolboxes and what they can do. As a result, their abilities sometimes seem to us to verge on the supernatural.

Consider some fascinating recent findings that received a good deal of attention from the media because they seemed to support the old and discredited idea of the inheritance of acquired characters. Several experiments were involved, some carried out by John Cairns and his co-workers at Harvard and others by Barry Hall, now at the University of Rochester. Some of the results are still subject to controversy, but some appear to be remarkably clear-cut.

Barry Hall's experiment is one of these latter. He found a system in which mutations apparently appeared only when they were needed rather than at random.

This is a high-tech story, but it starts with willow bark. For centuries, taken as a boiled infusion, this marvelous folk remedy seemed to cure practically everything. We now know that it contains salicylic acid, from which aspirin is derived. It also contains the beta-glucoside salicin, which in the last century was used as an anti-inflammatory agent and a general pick-me-up. Salicin can be split by the appropriate enzyme to yield glucose and saligenin, a phenolic alcohol.

All this is pretty arcane. The thing about this obscure compound that interested Barry Hall is that *E. coli* can use salicin—or more properly, the glucose it can obtain from it—in order to grow. But it does it in a very odd way.

Wild-type *E. coli* cannot use salicin. But mutants can. The mutants occur in a regulatory region that, once mutated, can turn on the enzymes needed to use salicin. The genes for these enzymes are there all the time, but the wild-type *E. coli* cannot access them; they are *cryptic*. The mutations needed to turn on these cryptic genes can either be point mutations or insertions of mobile elements.

Of course, *E. coli* can only turn the cryptic genes on and utilize salicin if the cryptic genes are functional. If they are not, then the cell is two steps away from being able to utilize this compound. It must turn the cryptic genes on, and it must fix whatever is wrong with the cryptic genes.

This is the situation Barry investigated. He found a strain with an IS element inserted into one of the cryptic genes. Now, in order to use salicin, the cell would have to turn on the cryptic genes *and* remove the IS element so that the cryptic genes could function.

He carefully measured the spontaneous mutation rates for these two events. A spontaneous mutation that turned on the cryptic genes was found to occur in only one out of every twenty million cells. And the spontaneous removal rate of the IS sequence from the damaged cryptic gene was even lower, so low that he could only estimate that it took place in less than one out of every five billion cells.

If mutations of both these types are truly random, a double mutation should occur in less than 1 out of 10^{17} cells. To find such a mutant, he would have to grow about a cubic yard of *E. coli*.

In fact, when he plated out these cells on a medium containing salicin and some other compounds that allowed them to grow for a few generations, he found that they made only small colonies because they could use the other compounds but not the salicin. But after a couple of weeks he examined the colonies again and found that about two-thirds of them showed wartlike growths that turned out to be made up of cells that *could* utilize salicin. The impossible had occurred— and at a rate at least a billion times higher than would have taken place had the mutations occurred randomly! Further investigation showed that the mutation that activated the cryptic genes occurred first, at a rate not very different from the rate he had measured earlier. What had increased mightily was the rate of excision of the IS element from the cryptic gene.

We have already met many situations in which unexpectedly mild environmental stimuli cause jumping genes to move (some indeed are so mild that we have not yet been able to determine what they are). But this case was different because he was able to show that it was the addition of salicin itself that increased the mutation rate. If salicin was left out of the medium, no excision mutants appeared in the little colonies. And with or without salicin, the mutation rate at other genes remained the same. Aha, trumpeted the media when they got hold of the story, the inheritance of acquired characters!

Well, not quite. What Cairns and Hall had shown was that certain environmental conditions could greatly increase the rate of specific mutations. The fact that the environmental conditions were related to the mutants could have been a complete coincidence, or could have been a reflection of the evolutionary history of this particular system.

The questions raised by these experiments leap out at us. The presence of salicin in the medium appears to trigger the excision of a specific element from the cryptic gene in the salicin utilization pathway. Do similar compounds trigger this event as well? Have similar situations arisen so often in the past that the likelihood of such an event has been greatly increased, perhaps by the interaction of salicin or similar compounds with specific cellular proteins that trigger the event? Can mutants be made that no longer trigger this event, or that cause it to happen even in the absence of salicin? If so, then it should be possible to probe the precise biochemical events that produce this phenomenon.

All these possibilities are under active investigation. This line of research is very exciting because, like so many other experiments we have talked about, it raises fundamental questions about the processes by which mutations arise. And the most likely interpretation of the experiments of Cairns and of Hall is that they have revealed the process of evolutionary facilitation. Certain evolutionary directions have been made very easy, and sometimes these evolutionary directions can be triggered by the very environment that the organism must adapt to. This can only have happened, not by some magic, but because during its long past history the organism has built the appropriate evolutionary toolbox.

SEX AND THE MOBILE ELEMENT

One of the great advantages of jumping genes and other mobile elements is that they can be harnessed to carry other genes that geneticists are interested in. They make excellent tools for transferring genes between different strains of bacteria and even between different bacterial species.

So you can imagine the excitement when it was discovered by Gerald Rubin and his colleagues at Berkeley that mobile elements could transfer genes between different strains of Drosophila. This opened up a world of possibilities. Certainly, it is child's play these days to put genes from Drosophila into bacteria. But when this is done, the genes usually either do not work or do not work properly. The ones that can be made to work after some modification are the common or garden variety of genes that produce specific proteins.

The most interesting genes, such as those concerned with development, are quite unable to perform their developmental job in bacteria even if they can be persuaded to produce proteins. They cannot make a bacterium grow legs or wings. But now, as we saw in the last chapter, it has become possible to take a fly gene, modify it, and stick it back in a fly to see what happens.

Already, such experiments are telling us a great deal about jumping genes and the evolutionary toolboxes of which they are a part. To understand these experiments, we need two pieces of information.

The first is this. One major reason that fly genes will not work in bacteria is that they have introns, while those of bacteria usually do not. You will recall that introns are pieces of DNA, inserted at one or more places in the gene, that interrupt the coding sequence. After messenger RNA is made from the gene, the introns must be removed from this RNA before it can function to make the right protein. This happens in the nucleus, and it is usually carried out by small nuclear ribonucleoproteins, or *snurps* for short. The snurps act as enzymes to remove the introns with great precision.

Genes from which the introns have been removed by the investigator can also be made to work in Drosophila by inserting them in a mobile element and microinjecting a small quantity of this mobile element into the developing egg.

The second bit of information we need to know is how the Drosophila mobile elements were discovered. Unlike most jumping genes, they were originally detected because they appeared to have evolutionary effects.

Laboratory strains of *Drosophila melanogaster* are carefully nurtured and have lived off the fat of the land for thousands of generations. Other members of this same species, of course, still live in the wild. It was observed by several groups of workers that strange things happened when their laboratory strains were crossed with flies of the same species caught in the wild. The mutation rate went up dramatically. Hybrid female flies showed varying degrees of sterility. Chromosomes broke and rearranged themselves at a much higher frequency than normal. The effects were short-lived, usually vanishing after the first generation. And they normally only happened when wild males were mated with laboratory females, not the other way around.

Because the hybrids suffered in so many ways, the phenomenon was named hybrid dysgenesis. Population geneticists suspected that it might be a mechanism for producing new species because it has the effect of isolating wild and laboratory lines of Drosophila from each other to some degree, even though they are the same species. The speculations about speciation have died down somewhat, in part because the phenomenon disappears quickly and only seems to happen under unusual conditions. But it was not long before the molecular mechanisms underlying hybrid dysgenesis came to be understood, at least in part.

One kind of hybrid dysgenesis (there are at least two) is caused by the shifting within the genome of transposable factors called P elements. The elements vary in length. A fully functional P element is about three thousand bases long, and it codes for a protein called a transposase, which is necessary for it to make copies of itself. The copies can insert themselves all over the genome, though they tend to settle in certain places. Like bacteriophage Mu, they often make mutations where they settle. And they can even break chromosomes, though the mechanisms by which they do so are obscure.

A working P element cranking out transposase somewhere in the genome, can also cause little stumps of P elements, dozens of which are usually lurking around the chromosomes, to start jumping too. All this frenetic activity can actually produce so much damage that it leads to a reduction in fertility. This is because many of a female's eggs may carry new lethal mutations or suffer chromosomal breaks.

Complications abound in the P system. Wild flies have thirty to fifty P elements, while domesticated flies have none. You might think that the wild flies would be in genetic trouble, with elements leaping about each generation. But in fact a mass movement of these P elements will not be triggered until a male carrying P elements mates with a female carrying none. This phenomenon is not yet understood. The result, though, is that in wild lines of flies the P elements stay put and behave themselves. At least they do not jump about very often, though, as we will see, they can move.

Further, even when the P elements are triggered and jump about like peas on a griddle (no, that's not why they're called P elements),* they jump about only in the sex cells and not in the somatic tissue. Some brilliant recent work, again from Rubin's laboratory, has revealed in remarkable detail why this happens. There are three introns that break up the transposase gene and that have to be chopped out of the RNA message before it can make the transposase protein. Molecular detective work showed that only two of these are removed in the nuclei of cells of the somatic tissue—as a result the transposase cannot be made. It can be inferred that the only place all three are removed is in the germ line cells that produce the gametes. This explains why P element transposition is confined to the germ line.

This could not be checked directly, because it was not possible to isolate enough germ-line cells to see whether the third intron was removed. Instead, a beautiful experiment was done in which a P element lacking the first two introns was cloned. The third intron was then carefully removed from the cloned element. When this modified P element was reintroduced into a fly, it transposed everywhere in the fly, not just in the germ line. The experiment could be arranged in such a way that when the P elements jumped in the somatic cells, little patches of mutant tissue appeared on the fly, the fruit fly equivalent of the red spots on the snapdragon flower. Figure 9.3 shows the effect on flies of the removal of the third intron.

Rubin's group suggested a good reason why the activities of P elements are confined to the germ line and to certain types of crosses. If P elements were allowed to transpose every generation, and in the somatic tissue as well, they would cause so much damage to the fly as it develops that it might not survive. This makes sense. But it also implies that

---

* The term comes from P and M, the paternal and maternal lines of flies that produce the hybrid dysgenesis effect when they are mated to each other.

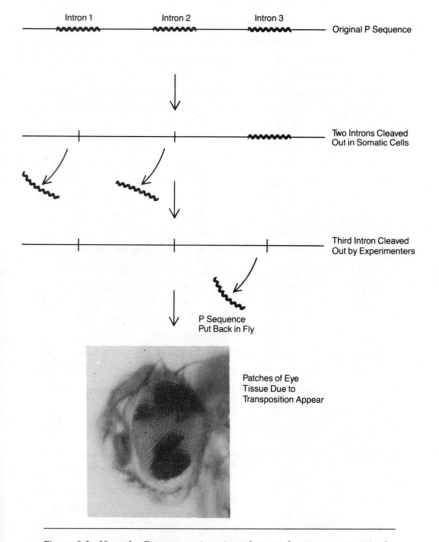

Intron 1   Intron 2   Intron 3

〜〜〜   〜〜〜   〜〜〜   Original P Sequence

Two Introns Cleaved
Out in Somatic Cells

Third Intron Cleaved
Out by Experimenters

P Sequence
Put Back in Fly

Patches of Eye
Tissue Due to
Transposition Appear

Figure 9.3 How the P sequence is activated to produce transposase in the germ line of Drosophila. Two introns are cleaved out in the somatic cells, but the third, which permits the transposase to be made, is only cleaved out in the germ line. When the third intron was removed by the experimenters and the completely processed sequence put back in the fly, it caused transpositions in the somatic cells. The picture shows the results of one of the genetic tests used, which produces patches of pigmented cells in the eye. (From figure 6 of F. A. Laski, D. C. Rio, and G. M. Rubin, Tissue specificity of Drosophila P element transposition is regulated at the level of mRNA splicing, *Cell* 44[1986]:7–19.) Copyright © 1986 by Cell Press.

there are times when P elements *do* transpose in wild populations. This is because these events must happen often enough to put a selective premium on confining their activities to the germ line.

The factors that trigger such transpositions are as yet mysterious. In 1987 Christian Biémont and his colleagues at the Université Claude Bernard reported on such an event in one of a number of inbred lines of Drosophila that they were checking periodically. Somewhere between generations 52 and 69 almost all the *copia* elements in this line took it into their heads to shift about to new chromosomal locations. The other lines, treated (so far as they could tell) in exactly the same way, showed the occasional shift of one of the elements, but not this massive effect. P elements and other mobile elements stayed put. To use a terminology current among my students, this event was spooky.

Questions abound. Are such mass transpositions triggered by some unusual environmental circumstance? It need not be a violent one if the French workers' observation is any guide. Is there some optimal number of transposable elements, or does this vary depending on the environment? And why have domesticated lines of flies lost all their P elements, including the little stumps?

One consequence of this loss of P elements is that laboratory flies may have lost some potential for generating genetic variability. Do other domesticated organisms lose their jumping genes over time? Has it happened to us? Did our Paleolithic ancestors have more jumping genes than we do, and did we lose them when we invented agriculture and civilization? What fun to find out whether the !Kung of the Kalahari or the Australian aboriginals have more mobile elements in their genomes than effete Western man.

The P elements of Drosophila, in spite of the restrictions that the flies have managed to place on their activities, seem to be very like Mu in their effects. We have yet to discover anything as potentially destructive and creative as these elements in our own genomes, though a growing number of mammalian mutations have been traced to the insertion of viral chromosomes or parts of viral chromosomes into the mutant genes.

OPENING THE P ELEMENT TOOLBOX

Most of the easily detectable consequences of the movements of P elements seem to be deleterious. But it may be that they can produce other kinds of genetic variation—variation that might lead to increased fitness or to a more rapid adaptation to a changing environment. Experiments designed to try to detect such an effect have given tantalizing results, but they are difficult to interpret.

Trudy McKay of the University of Edinburgh has reported on some thought-provoking selection experiments in which she has used P elements to explore this possibility. The experiments involved artificial selection carried out in the laboratory.

One particularly popular character to select for, both because it is easy to measure and because it responds well to selection, is the number of abdominal bristles. These little bristles are found on plates on each segment of the abdomen of the fly, and a wild-type fly usually has about twenty of them per segment. They probably aid in preventing the fly from being wetted, but whatever their function there is a good deal of variation in number of bristles from fly to fly. Patient selection for flies with fewer or more bristles for a couple of dozen generations will produce lines of flies with only four or five bristles per segment, and other lines that average forty or more. It is impossible to avoid some inbreeding during the selection process, and because of this, the flies are often not as vigorous as the lines the experimenter started with. Aside from this and the altered bristle number, however, they appear exactly like the flies at the beginning of the experiment.

Trudy McKay set up two selection experiments. In one, she began by mating males with P elements to females without P elements. This permitted the P elements to move about during the formation of the next generation, although in subsequent generations they were supposedly confined to quarters again. In the other experiment she started with exactly the same types of flies, but began the selection with matings in which females with P elements were crossed with males lacking them. In this line, the P elements were never allowed to move.

The results were remarkable. The selective response, that is, the rapidity with which bristles were gained or lost, was over twice as great in the line in which the P elements had been permitted one generation of movement.

Subsequent experiments by her and by several other workers showed that the effect was not as pronounced as it first appeared. This may in part be due to the phenomenon, known to all scientists, that a new experiment will often give superb results the first time and never give such clear results again. It also appears that the P elements were not as imprisoned as she had thought they were in her control lines. This had the effect of blurring the differences between the control and experimental flies. Her later experiments showed that variation of several kinds could accumulate more quickly in flies carrying the P elements than in those without them, even when the elements would theoretically not be expected to move. Presumably they were moving, but at a much lower rate.

Despite the difficulties in interpretation, these experiments showed that P elements really are important in generating the kind of genetic variation that can bring about phenotypic alterations in the properties of flies. The effect of their movement is not simply confined to sterility, point mutations, and chromosomal breaks; nor are their movements confined to crosses between lines with and without P elements.

Further, P elements appear to be extremely expert at producing such selectable variation. It is astonishing that a few dozen P elements can produce so many mutations in the genes responsible for determining the numbers of abdominal bristles. Perhaps the P elements preferentially affect a small number of genes that are very important in development, and this has a kind of cascade effect so that each change affects many different pathways. McKay's results seem to tell us that the P elements and the genes they affect can work together in very specific ways to produce selectable genetic variation.

Regardless of the explanation, it appears that the harmful effects we see, the ones that initially alerted geneticists to the existence of the P elements, are only the tip of the iceberg. The major effects of P elements may only be detectable by experiments like McKay's. The next step is to trace where the elements go and then to clone their favorite settling places. P elements provide a means to clone genes, such as those responsible for tiny variations in the appearance of the fly, that are not approachable by any other method.

These few stories are a fair sampling of the many that are appearing as a result of the meshing of genetics, evolutionary theory, and molecular biology. The tens of thousands of scientists working in these areas are already coming up with many more.

All the stories show the emerging importance to evolution of organized genes, gene complexes, and mobile elements. Other patterns can be seen to emerge as well. One of the most striking is that different evolutionary toolboxes show different degrees of sophistication, as measured by the relative amounts of random and less random change that they introduce into the genome. Another is that organisms, particularly higher organisms, have thrown up strong defenses against these changes, sometimes preventing them from occurring and sometimes confining them to particular areas.

But how can these toolboxes have evolved? This brings us back to the nagging question posed at the beginning of the book. Most of the time, most organisms are well adapted to their environments, a world full of Babbitts. What possible advantage is there to such individuals to possess evolutionary toolboxes that can lead to dramatic mutations? Look what happened to Babbitt when he began to stray.

10

Potential-Realizing and Potential-Altering Mutations

The story has been told about a visit Henry Ford made to the Stanley [Steamer] plant in 1905. Ford asked the brothers how many cars they produced in a year. The Stanleys said 650, to which Ford replied, "Why, we make that many in one day in my factory."

—ANDREW JAMISON, *The Steam-Powered Automobile*

ANOTHER AUTOMOTIVE EXCURSION

In January 1907 a specially designed Stanley Steamer automobile took part in speed trials on the beach near Jacksonville, Florida. It was driven by Fred Marriott, the Stanley brothers' race driver and mechanic. The previous year, the car had been clocked over a measured distance at 127.5 miles an hour, breaking by 20 miles an hour the previous landspeed record, which had been set by a gasoline-powered Mercedes-Benz.

F. E. and F. O. Stanley were identical twins* who brought

---

* So contemporaries agree, though no blood tests were done.

mechanical ingenuity and conservative New England marketing methods to the building of steam cars. They stood directly athwart the rising tide of internal-combustion engines, and throughout the lifetime of their company they managed to sell virtually every car they made. The Stanley cars were quiet, powerful, and gearless, though not particularly energy efficient—they burned at least twice as much fuel as their noisy, gear-clashing, gasoline-powered opposite numbers, although the fuel was cheaper. They also cost five times as much as a mass-produced Ford, the car that was already on its way to transforming the American landscape.

By 1907 steam had been proving itself for well over a century. The first steam-driven conveyance, designed as an artillery truck, had been built by Nicholas-Joseph Cugnot in 1770. It was capable of carrying four people at two and a half miles an hour. Throughout the nineteenth century, steam vehicles for public transport developed in parallel with the railroads, although the railroad interests did their best to suppress them.

Internal-combustion engines were parvenus by comparison. Karl Benz's first cars appeared on the market in 1890, at about the same time that Léon Serpollet made the first practical steam cars for individual owners. Internal combustion quickly prevailed, however. By 1907 it was well on its way to winning the battle. Nonetheless, the Stanleys' future sales and expansion plans depended on the headline-grabbing potential provided by a new land-speed record.

Their car, known as the Wogglebug, approached the beginning of the measured mile at tremendous speed. When Marriott crossed the line, the car was no more than a blur. F. E. started his stopwatch when he saw the puff of smoke from the electrically triggered signal gun.

Accounts of what happened next vary, but all agree that a few seconds after the start of the run the car hit a series of ridges on the beach, the front wheels lifted off the sand, and the whole car planed through the air for a considerable distance before crashing to the ground and dissolving in a welter of parts. F. E. stopped his watch by reflex when the car started to veer, so he was later able to calculate that it was traveling at about 150 miles an hour when it was turned briefly and involuntarily into an airplane.

Though badly injured, Marriott eventually made a complete recovery. The accident so unnerved the Stanleys, however, that they withdrew their cars from the racing and speed trial circuit where so much publicity could have been gained.

Stanley Steamers continued to be made in various guises through the 1920s, although F. O. Stanley sold the company after his brother was killed in a motoring accident in 1917. After that, the few steam cars that were built were expensive handcrafted curiosities appealing only to an ever-diminishing number of steam aficionados. The moment of steam had passed, defeated by the Stanleys' innate conservatism and Henry Ford's invention of the assembly line.

It is useless to speculate about what might have happened had the Stanley brothers invented the assembly line rather than Ford. Their cars, while having many advantages, were monstrously difficult to start. Unlike Serpollet's much earlier and safer cars, they had a boiler rather than a flash-type generator for producing steam. This meant that they were much more likely to explode, hardly a selling point. These problems were in principle fixable, and indeed Abner Doble's elegant steam cars of the 1930s started as easily and were as safe and responsive as any other car on the road at the time. Still, had the Stanleys won, the world might have been a very different place, with gentle steam-kettle-like hissings replacing the throaty roars of the internal-combustion engines that now assault our ears.

The moral of this tale is that both the organization of the factory and the ability to make small but important changes have been essential to the success of the internal-combustion automobile. Though driven by natural selection rather than market demand, exactly the same factors operate to determine the success of living organisms.

MUTATIONS THAT CHANGE
EVOLUTIONARY POTENTIAL

I began this book by reaffirming that the basic processes of evolution as we currently understand them are quite enough to account for the diversity of organisms in the world around us. But I also proposed that arising from these processes there is a kind of higher-order structure to evolution, a structure that facilitates the evolutionary process itself. We have by now looked at enough different examples of this higher-order evolution to be able to see in outline how it might work. Let me expand on the metaphor of Ford's invention of the assembly line to propose a simple model. The model is completely consistent with known evolutionary theory but explains how genes can have become well organized over time and can in some cases even appear to anticipate future environmental changes.

In the last few chapters we have opened and examined some of the evolutionary toolboxes available to organisms. Many of these toolboxes are supergenes, tightly grouped and organized collections of genes that govern more than one cellular or developmental function. Other toolboxes have the capability of shuffling genes and bits of genes and putting them together into new and potentially useful combinations. The existence of these toolboxes changes the *potential* for evolution.

We can now see that the toolboxes themselves are evolving in two different ways. Each of these types of evolutionary change proceeds at a different pace.

The fastest and most common kind of change is produced

by mutations that alter the immediate function of the tool-boxes, or genes on which the toolboxes act, without altering their integrity or organization. These have their equivalent in the world of automobiles in the form of small yearly design or engineering changes that may enhance or diminish the sales appeal of the cars. They can be accomplished without altering in any fundamental way the methods by which the cars are made.

In the world of nature, typical examples include the somatic mutations that fine-tune the antibody genes, or the mutations that alter the shape or position of spots on a butterfly's wing. Others are small alterations in the amino acid sequences of the glycolytic enzymes, which may enhance or diminish their function. The survival or disappearance of this kind of mutation is governed by individual selection, for the impact of these changes is felt immediately by the individuals that carry them. We can call these *potential-realizing* mutations, because they realize or implement the potential of the genetic toolboxes.

The second type of change occurs less often. These are mutations that change the organization of the toolboxes themselves. Let us call these *potential-altering* mutations.

Henry Ford's reorganization of his factory was certainly a fine example of a potential-altering mutation in the automotive world. In the antibody system, such mutations include increases or decreases in the number of V or C genes, and shifts in their position so as to alter the likelihood that particular classes of V or C genes will join together. In the butterfly supergenes, they include the insertion of new behavior or morphology genes into the complex, giving the potential of greater fidelity of the mimic to the model.

Potential-altering mutations also include subtler changes. For instance, the DNA sequence of a region flanking an important enzyme domain might be altered, increasing or decreasing the likelihood that this domain will be moved elsewhere in the genome by means of the action of jumping genes.

These potential-altering mutations may, like potential-realizing ones, confer immediate benefit or harm on the organism carrying them. But this is their least interesting property. What is much more interesting is that they can alter the potential for evolution at some point or points in the future. Because of this, the survival or disappearance of potential-altering mutants is governed by more complicated factors than individual selection.

The management of an automobile company may decide to reorganize its factory, à la Ford. They take a chance in doing so, for this change might in the short term be harmful—the factory may have to be shut down for a time, or financing of the alterations may be a strain. The management hopes that after the reorganization it may be possible to implement much more sweeping or rapid changes in its cars in order to take advantage of changing customer requirements. But just as in the natural world, potential-altering changes that are made in the organization of a factory might be harmful farther down the line, even if not at first. Robots may be brought in to replace human workers, triggering a wave of strikes—and then the robots may be unable to do the job after all. Fancy executive suites might be built, only to be staffed with executives who spend their time drawing huge salaries and giving each other lunch. As with potential-realizing mutations, a potential-altering mutation may be harmful rather than beneficial. Unlike potential-realizing mutations, however, this harm or benefit may become apparent only after a period of time.

The lapse of time may in theory be short or long. New potential-realizing mutations will begin to arise immediately after the appearance of the potential-altering mutation. This might allow a population to adapt quickly to changing conditions. For example, a potential-altering mutation occurring in a butterfly mimicry supergene complex might allow that population to adapt quickly to the appearance of a new model. This would happen because the potential-altering mutation enhances the likelihood that advantageous potential-realizing mutations will appear subsequently. Potential-altering mutations of this type can easily be selected for because their advantage quickly becomes apparent.

In other cases, the true adaptive advantage or disadvantage of a potential-altering mutation might only become apparent at a much more distant time. During a wave of extinction, conditions change rapidly. Potential-realizing mutations facilitated by a potential-altering mutation that took place many generations ago might suddenly become advantageous and perhaps save the species carrying this mutation from extinction.

I think that this second kind of situation can happen only rarely. It would require a potential-altering mutation to arise, survive, and spread through a population by chance even though its immediate effect and the effect of the mutations it facilitates might be harmful in the short term. It is as if the Stanley Steamer Company had somehow been able to survive all the decades during which internal-combustion cars have dominated the world. If the world then ran low on gasoline, the possession of factories capable of producing steam automobiles could in theory catapult the Stanley Steamer Company into world prominence. But this cannot happen because that whole manufacturing toolbox was lost decades ago.

In order to survive, most potential-altering mutations must have an advantageous effect on the organisms that carry them, and that effect should be expressed within a relatively few generations. Otherwise, like most mutations of any type, they will almost certainly be lost.

PREPARATION FOR THE ICE AGES

We can see the interplay between potential-realizing and potential-altering mutations at many levels. Consider the mammal populations of Europe at the end of the Pliocene, about two million years ago. Conditions were changing for the worse at that time, with the first Pleistocene ice age soon to come. There had been a gradual cooling trend throughout the Pliocene and the preceding Miocene. The flooding of the Mediterranean basin some three million years earlier, surely one of the most spectacular events in geological history, had effectively cut off the possibility that the mammals of Europe could escape to the milder climate of Africa to the south. They were forced to face the coming ice ages without the option of retreat.

In spite of these portents of trouble, things were looking pretty good in the short term. At no point in the history of the mammals, before or since, have there been so many species.

A visit to the European or Asian grasslands of the Pliocene would remind you irresistibly of today's East African savannahs. Vast herds of antelopes and gazelles, quite familiar in appearance, migrated for long distances in search of fresh grazing just as African ungulates do today. There were smaller herds of several species of elephant, rhinoceros, and giraffe. Horses and camels were also present, though in fairly small numbers, migrants from their evolutionary home in North America. These grazing animals were preyed upon by a variety of species of large cat, including sabertooths. The parts left over were devoured by spotted and striped hyenas.

Hippopotamuses wallowed in the slow-moving rivers that wended across the European plains. Many species of wild pig, of whom the wild boar is now the only remnant in Europe, flourished in the upland forests. The mild climate, luxurious vegetation, and varied landscape provided an abundance of different ecological niches that encouraged speciation. The amazing variety of animals we see in East Africa at the present time must be only a pale shadow of those that existed in the Pliocene.

But things were soon to change. A series of severe glaciations were to drive many species to extinction. This wave of extinctions was accompanied by the appearance of a much less numerous set of species adapted to the severe glacial climate of the Pleistocene. The hippos and giraffes of Europe disappeared, as did the tropical rhinos. The latter were replaced by a new species of rhinoceros with a coat of woolly hair that enabled it to survive down to quite recent times. (It was man who probably finally did in the European rhinoceros, after painting its portrait on the walls of his caves.)

The new species arose from a small subset of the original Pliocene set of species. How did this lucky subset differ from the others? One explanation, consistent with the suggestions I have made so far, is that some of the Pliocene species had a slightly different genetic architecture from the others, an architecture that enabled mutations leading to cold-weather adaptation to arise in them more readily.

Many cold-weather adaptations did take place. They included longer hair, accompanied by an increase in the thickness of the underlayer of soft, curly barbed hairs that provides most of the insulation. The sizes of the ears and other exposed parts were reduced, and changes in blood flow and thermoregulation cut down the amount of heat lost through the surface of the body. Behavioral changes in small mammals included an increased tendency to store food and a switch in dietary preferences toward foods that lasted longer when stored. For large mammals, burrow building and hibernation were obvious adaptations.

None of these changes was particularly dramatic, but taken together they added up to survival rather than extinction. I am sure that any one of our Pliocene species would have been able to evolve these changes given enough time. But obviously those that did so first had the advantage. Does this imply that the adaptation to a glacial climate took place as a result of species selection? Was it simply competition between species that could by chance adapt more or less quickly to the cold? Or were the evolutionary events at the end of the Pliocene more complicated than that?

This brings us back full circle to the nagging problem of species selection versus individual selection. Suppose that those animals that won the late Pliocene evolutionary race did so because they had a genetic architecture, and an accompanying complement of mutagenic viruses and transposable elements, that enabled them to be the first to adapt to these new climatic extremes. This would be the result of one or more potential-altering mutations. If there had not previously been selection for this new genetic architecture (since glaciations had not happened for many millions of years), then these new arrangements must have arisen simply by chance. As a result they would not be very different from the old genetic arrangement, for if they were very different they would probably have had pronouncedly harmful effects when they first appeared. Further, they would tend to facilitate the production of many potential-realizing mutations that would not be advantageous during an interglacial period.

Selection for such a new and slightly different genetic architecture would, in the species selection model, happen only with the onset of glaciation. If evolutionary facilitation was forced to evolve in this slow and intermittent fashion, it could hardly have had much impact on the organization of the genome.

There are two hidden assumptions in the scenario I have just laid out for you, and both of them are wrong.

The first assumption is that adaptation to new conditions, such as a period of glaciation, involves selection for mutations and gene combinations that are different from any that have been seen before. But this is certainly not true. Adaptations to cold climate and severe winters must have occurred repeatedly in Pliocene Europe, though perhaps only in localized areas. Further, the kinds of adaptations to the onset of glaciation that I listed as taking place in our Pliocene mammal populations were not qualitatively or quantitatively different from those that would adapt such populations to any cold climatic regime. And for the most part the adaptations were rather simple, involving an increase or decrease in the size of an organ, an increase in the numbers of hair follicles, and so on.

Similar adaptations, and their reverse, must have occurred repeatedly in the ancestries of the Pliocene animals. Indeed, it seems likely that those species best able to adapt to the onset of glaciation were those that had already repeatedly adapted to living at high altitudes or latitudes. Thus, the advantageousness of their potential-altering genes could have been tested soon after they arose.

I am tempted to compare these repeated rounds of adaptation to a pianist warming up with a few scales and finger exercises before a concert. Of course there is a difference, for the pianist knows when the concert is to start. The Pliocene mammals had no idea that glaciers would soon come crunching down from the north and the high mountains to overwhelm them. But, whether conscious or unconscious, the effect on survival or success is the same. You will remember the story of the tourist who asked an old man on the street how to get to Carnegie Hall. He replied, "Practice! Practice!"

The second assumption that must be challenged is that a given species is genetically uniform at any particular time. For example, all the members of species A might have a slightly different gene arrangement from all the members of species B as a result of a potential-altering mutation that has spread through all the members of species A. This in turn facilitates the appearance of selectively useful mutations in species A, so that it can evolve more rapidly when the environment changes. The idea of genetic uniformity is an important hidden assumption underlying the idea of species selection. Species are the units of species selection just as individuals are the units of individual selection, and since they are supposed to be units, they are tacitly assumed to be made up of individuals all having the same properties.

In order to destroy this assumption, I must begin with a historical digression.

From the very beginning of the science of genetics at the turn of the century, geneticists have looked for lines of laboratory animals and plants that are uniform and easy to work with. This was especially necessary at the outset, when the mechanisms of genetics were not fully understood. Genetic variability in characters other than those being studied did nothing but add confusion.

Laboratory organisms have been established through a lengthy process of selection for genetic uniformity. Much inbreeding has taken place, some deliberate and some by chance. The details of these procedures have often been forgotten. The laboratory lines of Drosophila that I mentioned earlier were established in just this fashion, and have in the process lost many of their jumping genes.

The result of all this conscious and unconscious selection was the establishment of laboratory strains that had no obvious mutational defect. These were called, with faint irony, the *wild type*. Because the wild type did not change with time, it became a reference point for the geneticist, who could always compare mutant strains to it. Geneticists have thus come to think of their experimental organisms as falling into one of two classes: wild type or mutant. And unfortunately, this dichotomous *typological* thinking (as the Harvard evolutionist Ernst Mayr termed it) has also come to pervade the way that many biologists think about evolution.

I do not want to belabor the point, but one egregious example will do. Before the advent of DNA sequencing methods it was, as I mentioned in chapter 6, an enormous task to determine the amino acid sequence of a protein. A proud protein chemist might, in many cases after years of work, finally achieve an unambiguous sequence, consisting of hundreds of amino acids, of a protein from a pig. This sequence became enshrined henceforth as the sequence of the pig. Molecular evolutionists then compared it with the sequence of the dog, the goat, the whale, and so on, each achieved by a similarly tedious effort.

It would of course have been considered the height of foolishness to go back and spend more time sequencing the same protein from another member of the same species. It would probably be the same anyway, or would differ at the most by a very few amino acids.

Even rather important distinctions tended to fall by the wayside as a result. Molecular evolutionists spoke about the sequence of the whale without bothering to find out from which species or indeed from which of the six different families of whales the protein had actually come. This attitude, that there was one "right" sequence for each species, inevitably led molecular evolutionists into thinking that the process of evolution is simply a replacement of the sequence of an ancestral species with a slightly different sequence in a descendant one. They did not concern themselves with the details of how this replacement took place.

By contrast, population geneticists are now beginning to use the techniques of molecular biology to measure the variation in DNA within a species, and they are finding lots of it. Indeed, it is possible to use these variations as markers for tracking down the location of many important human genes.

Unraveling the nature of these small changes is where the study of microevolution comes into its own. This is the domain in which selection on individuals becomes important. Paleontologists and molecular evolutionists either pay no attention to such piddling details or—to be unkind—have never learned about them in the first place. But most macroevolution is made up of innumerable microevolutionary steps.

We are easily able to perceive the diversity in appearance of members of the human species (which reflects an underlying genetic diversity). Yet most of us are struck by how remarkably similar sheep appear to be; only the shepherd will swear that all the sheep in his charge are distinctly different in personality, appearance, and behavior. Indeed, unless we are animal behaviorists, it is difficult to measure the true diversity of other species. But they are diverse, and rather than speaking of *the* pig or *the* whale, it is more appropriate to think of each pig and each whale as being clearly distinct from all other members of the same species.

Even a superficial similarity masks great genetic diversity. This is apparent at every level, from our superficial appearance and the proteins that make up our bodies down to the DNA that codes for them and the chromosomes in which the DNA resides. At the DNA level, there are literally millions of differences between any two human beings—unless, of course, they are identical twins. In the human species, the diversity within a racial group is far greater than the diversity between groups, a valuable reminder that racism is founded on ignorance.

This diversity, hidden in every species, applies to both potential-altering and potential-realizing mutations, with the result that many different such mutations may coexist in any population of a species at any one time, maintained by a complex interaction of selective pressures. You need only recall the story of the mutations conferring resistance to malaria in the human population to see how this could occur.

In the case of the Pliocene mammals, each population would have harbored a whole range of potential-altering mutations, giving rise in turn to a bewildering variety of potential-realizing mutations. Some of these potential-realizing mutations conferred resistance to cold weather, while others conferred resistance to hot or dry weather elsewhere in the species range.

Severe winters and invasions of high-altitude regions provided repeated though quite unconscious dress rehearsals for the main event of glaciation. But this was not the only thing going on in these populations. The glaciers might not have appeared, and the European plains might instead have been subjected to millions of years of a hot dry climate. Equally unconscious dress rehearsals for that possibility were going on at the same time, made possible by the pool of genetic variation within each species. This kind of evolution must have been more or less continuous and not dependent on the achingly slow process of species selection.

I cannot resist falling back on the automotive analogy one final time. There are hundreds of different kinds of capitalist cars on the market, made by dozens of factories with very different organizations. The cars and their factories are constantly changing. When the customers begin to worry about fuel shortages, tiny, fuel-efficient cars quickly appear. When, as now, we are swimming in a sea of oil (and choking on its products), giant limousines and vast station wagons make a comeback.

The world of the Communist automobile is very different, much closer to the idea of a monolithic uniform species that is prevalent among many molecular evolutionists. One or two models are made, and they may bear no relationship to market realities. (Each Russian car, for example, produces between five and ten times as much air pollution as the average Western car.) These models change very slowly, as do their factories.

I do not wish to appear to be mindlessly extolling the virtues of the free market, particularly since unbridled consumerism in the West appears to be bringing on a new interglacial period much more swiftly than anyone would have believed possible a few years ago. But I want to emphasize that diversity in the world of both automobiles and species can accelerate not just their evolution but also their very ability to evolve.

11

Closing the Circle

We have covered a lot of ground since I posed some evolutionary questions at the beginning of the book. Let me now suggest some answers to them that are consistent with the ideas of genetic toolboxes and of potential-altering and potential-realizing mutations.

Consider the matter of sudden bursts of evolution, of the type that we see among the Drosophila of Hawaii. The conventional evolutionary viewpoint is that there was a great diversity of unoccupied niches available in the relatively insect-free Hawaiian Islands. (Tourists may think that Hawaii is anything but insect-free, but the mosquitoes that greet them in such numbers were introduced by man. To get away from them, the tourist must simply go to an area where there are few introduced plants and consequently few introduced animals.) Confronted with this great diversity of different living spaces or niches, the ancestors of the Hawaiian Drosophila were able to evolve rapidly to fill them.

Adaptive radiation was not confined to the Drosophila. There are at least eleven hundred species of Lepidoptera in Hawaii, most of them tiny drab moths that to the untutored eye look very similar to each other. They are the result of, at most, five introductions of Lepidoptera to the islands. (Contrast the British Isles, with almost twenty times the area, which have only twenty-five hundred species of butterflies and moths.) Further, the moths of Hawaii can evolve with amazing speed. A number of species of the genus *Hedylepta*, endemic to the islands, have adapted to the banana plant as their sole source of food, even though bananas were introduced less than a thousand years ago with the arrival of the Polynesians. This adaptation has apparently occurred several times independently, since different species of banana-eating *Hedylepta* are found on the various islands.

That these groups of insects, confronted by the Hawaiian situation, are capable of rapid adaptive radiation should not surprise us. Their radiation has been triggered by the unusual circumstances on these remote islands, particularly the large number of niches made available by the paucity of other insects. The Drosophila of other oceanic islands, closer to centers of insect diversity, have been prevented from radiating appreciably.

But I would like to suggest that, while the basic idea of adaptive radiation is quite correct, there may be an extra dimension to it. That dimension can be supplied by the organization of the insects' genes and the kinds of mutational changes that might preferentially arise in them. If the appropriate evolutionary toolboxes are available, then the flies and moths are able to radiate so quickly because they have a long history of being confronted with similar situations and responding to them.

Evolutionary toolboxes, as I pointed out in chapter 9, can be more or less sophisticated, depending on how much evolution they themselves have gone through. I keep a box full of basic tools in a cupboard, and with them I can perform a variety of emergency operations on our house, ranging from replacing a tap on a sink to repairing loose baseboards. Occasionally, however, my fumbling attempts at repair lead to worse damage—pipes are knocked loose and produce floods in unexpected places, electrical outlets inexplicably lose power. The plumbers, carpenters, and electricians I call in to repair my unwitting damage do so with a few deft manipulations of the tools in their specialized and highly evolved toolboxes. But plumbers do not perform the jobs of electricians, and vice versa. That incautious blurring of roles is left to bumbling amateurs like me.

We have already seen a number of examples in the natural world in which specialized toolboxes limit as well as enhance evolution, while less sophisticated toolboxes might lead to more dramatic alterations at the expense of a great deal of evolutionary garbage. If toolboxes exist that facilitate adaptive radiation in the Hawaiian insects, they must be quite specialized ones. This is because, in spite of all their adaptive radiation, the flies and moths of Hawaii are all still flies and moths. Many Drosophila have taken up a beetlelike existence in the underbrush and leaf litter of the native vegetation, but they have clearly not yet converged morphologically on beetles. It is possible that certain kinds of behavioral and morphological changes can occur easily, but others are much more difficult to accomplish. We will see in a moment some ways in which this possibility might be tested.

Another aspect of evolutionary facilitation is illustrated by the indigenous mammals of South America. These have often evolved in remarkable parallel with mammals elsewhere in the world. A single-toed foot is an obvious adaptation for running across open grasslands, and very large upper canines provide an obvious advantage for seizing prey. When these adaptations arose independently in South American mammals and in mammals elsewhere, did they do so because similar adaptive problems could only be solved in a restricted number of ways, or because certain evolutionary directions were more likely than others, or both?

It seems probable that evolutionary facilitation played an important role in these cases of parallel evolution. This is because the mammals of South America and those of the rest of the world had a very long common ancestry, much longer than you might suppose. During that long period, similar adaptations involving changes in the size of the toes or the pattern of dentition undoubtedly arose many times.

The eventful history of the mammals did not begin with the demise of the dinosaurs. It is true that mammals underwent a remarkable adaptive radiation into the niches that the dinosaurs vacated. But long before that, during all the time that the dinosaurs ruled the planet, small mammals were present too. True mammals probably appeared about 190 million years ago, a full 125 million years before the dawn of the so-called Age of Mammals.

Though the early mammals were much less diverse than they are now, a good deal of mammalian evolution went on even during the dinosaur era. We know of at least three distinct types of small mammals that lived during that time. Most of them were probably tiny insectivores that led very retiring lives, beneath the notice of the giant reptiles. The marsupial and placental lineages probably diverged about halfway through the Age of Reptiles, though there is much uncertainty about just when this split occurred.

Yet, if we follow these sparse traces back even further in time, they suddenly become abundant and diverse. The ancestors of these mammalian groups were the highly successful therapsids. These creatures have not received all the publicity accorded to the dinosaurs, yet they ruled the planet during the fifty million years just before those noble beasts appeared. There are actually more therapsid than dinosaur remains in the fossil record, vividly attesting to their success.

Recent evidence suggests that there may have been not one but two great extinction events that shaped the history of the mammals. The first occurred about 180 million years ago and resulted in the replacement of the therapsids by the dinosaurs as the dominant large land animals. The second occurred sixty-five million years ago, when the ruling reptiles were replaced in their turn by the mammals, the remote descendants of the therapsids. By one of the ironies of fate, the mammalian lineage got a second chance.

The therapsids, of course, arose from an even older set of ancestors. These were the pelycosaurs, which flourished during the late Carboniferous and early Permian, almost 300 million years ago. At that distant time our ancestors were not at all mammal-like, for the pelycosaurs were the immediate descendants of the first amphibians (see figure 11.1 for a picture of some representative pelycosaurs).

Unlike the amphibians, the pelycosaurs lived their entire lives on land, and in that sense they had features in common with their remote mammalian descendants. Evidence from their bone structure, footprints, and the ratios of carnivores to herbivores all suggest that they were quite cold-blooded. Even so, it is certain that we share many genes and even gene complexes with the pelycosaurs. It is therefore a matter of definition whether the pelycosaurs and their successors the therapsids should be included in our mammal-like ancestry.

There is a great deal of argument about just how mammal-like the therapsids were. One aberrant offshoot of the mammals that still survives in Australia consists of the duckbill platypus and the spiny anteater, the only living representatives of the monotremes. Like reptiles, these creatures lay eggs; and also like reptiles, the males have internal testes and the females have a cloaca. When did they split off from the rest of the mammals? The molecular data are too sparse as yet to be certain, but it may have been as far back as the time of the later therapsids. If so, then those therapsids might have been very mammal-like indeed.

The name therapsid comes from the Greek for ''animal''

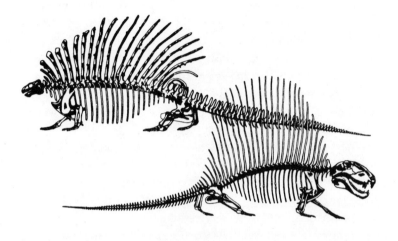

Figure 11.1 Typical pelycosaurs from the Pennsylvanian, just before the rise of the therapsids. Note the reptilelike stance, which did not permit walking or running for long periods. The one on the left is *Edaphosaurus;* the one on the right is *Dimetrodon.* The sail-like structure on the back may have been a primitive temperature-regulating device, or merely a means of making the creature appear larger and more threatening. (From figure 47 of E. H. Colbert, *Evolution of the vertebrates,* 3d ed. [New York: John Wiley and Sons, 1980].) Copyright © 1980 by John Wiley and Sons, Inc.

and "arch." If you place your fingertips at your temple and clench your jaw, you will feel muscles bunching. The masseter and temporal muscles lie beneath the cheekbone, joining the dentary bone of the lower jaw to the skull and providing powerful leverage. The development of an opening, or apse, at the temple in our remote ancestors allowed the muscles to attach more firmly and to bulge outward, which meant in turn that more massive muscles and more powerful jaws could evolve. The appearance of the arch was an enormous evolutionary advance: all the therapsids had it; the pelycosaurs did not.

You will recall the ancient Chinese curse: may you live in interesting times. The therapsids were certainly cursed in this sense, for they had to battle what is probably the greatest upheaval ever to hit the planet at the end of the Permian about 230 million years ago. This event, which certainly involved widespread glaciation, resulted in the extinction of over 95 percent of the species of marine animals. Many therapsids also went extinct at that time, but a number of lineages survived. Some of those that did were pushed much closer to a mammalian anatomy and physiology.

What were these creatures like? John C. McLaughlin, an artist turned paleontologist, has attempted to reconstruct some of them. While some of his reconstructions are open to argument (*everything* is open to argument in this field!), he has performed a valuable service by gathering together scattered information about their fossil remains and showing that the therapsids were quite an advanced collection of animals (see figure 11.2).

There were three distinct adaptive radiations of therapsids, two before the great extinction at the end of the Permian and one immediately after. The earlier radiations produced an astonishing variety of very unmammal-like creatures with names like gorgonopsians and bauriamorphs, including a huge group, the dicynodonts, that successfully survived the Permian-Triassic transition.

The dicynodonts and some of the gorgonopsians were heavy and lumbering, with two immense teeth in their upper jaws. They were adapted to a way of life that was enormously successful at the time but was eventually lost through the rise of the dinosaurs and the flowering plants. The dicynodonts' heavily armored fossils are plentiful, suggesting that they must have moved in giant herds through the Triassic landscape. Their dentition shows that they were capable of uprooting or biting through entire seed ferns, rather like those machines that munch through our southern pine forests today, converting trees almost instantly to pulp. The nutritional value of such a fare must have been very low, so they had

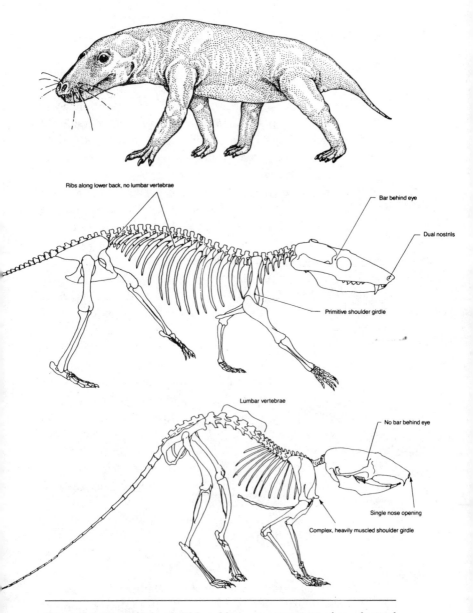

Figure 11.2 One of John C. McLaughlin's reconstructions of an advanced therapsid, *Thrinaxodon*, with vibrissae (whiskers). Below are pictures of an advanced therapsid skeleton compared with that of a mammal. Note that the therapsid has less-differentiated teeth, more ribs, and a primitive shoulder girdle leading to a stiffer and less flexible body—but that the overall stance and function of various parts of the skeleton are remarkably similar to that of the mammals that followed. (Redrawn from figures on p. 81 and pp. 122–23 of John C. McLaughlin, *Synapsida: A new look into the origin of mammals* [New York: Viking, 1980]).

to ingest enormous amounts of material. No moaning about too little fiber in the diet could be heard from these creatures.

Like their pelycosaur ancestors, the dicynodonts had the typical amphibian pelvic and pectoral structures, with the front and rear legs held at right angles to the body. Like today's lizards, they were probably able to move swiftly for short distances but would tire easily.

Other groups of therapsids that survived the transition to the Triassic, including the cynodonts or dogtooth animals, had a much more mammal-like skeletal arrangement, enabling them to run or walk for long distances. Many of them also had enlarged canines or caninelike teeth. And a number of lines evolved other mammal-like adaptations, so that by the middle of the Triassic—even before the rise of the dinosaurs—they were difficult to tell apart from primitive mammals.

Teeth make excellent fossils, so we know quite a lot about therapsid teeth. Many therapsids, particularly the later ones, had differentiated, mammal-like teeth that were adapted for grasping, shearing, and grinding. They arose independently in a number of therapsid lineages and were an enormous advance over the numerous but undifferentiated teeth of their ancestors. What the therapsids did not have until very late, when a few of these diverse lineages were shading into mammals, was any sign of a set of the deciduous milk teeth so characteristic of mammalian dentition.

Several lineages of therapsid took full advantage of their complex dentition by developing hard palates, literally a roof to the mouth. This separated the breathing passages from the mouth. Therapsids could breathe with their mouths full.

Most reptiles today cannot do this. When a boa constrictor catches a goat, it must first stop it from struggling by looping it with slowly constricting coils. Then it must cease breathing during the complex swallowing operation, which may take an hour or more. For an animal with a low metabolic rate, this is no hardship. But a carnivorous therapsid with its much faster metabolism could catch its prey using those powerful jaw muscles and keep breathing during the subsequent struggle. Unlike the boa constrictor, it could hold the prey in its mouth and carry it long distances before eating it or perhaps sharing it with its young. And it could breathe while dismembering the prey with its complex and differentiated set of teeth.

This is not to say that the later therapsids actually did all these mammal-like things. But their skeletal features, including the mammal-like internal organization of their bones, suggest that these things were possible. It can at least be concluded that they were *potentially* mammal-like in their range of behaviors and abilities.

The skull of the small late therapsid insectivore *Thrinaxodon* had little pits on the snout area, very similar to the pits in the skulls of present-day mammals that accommodate the root bulbs of whiskers. If this creature had whiskers, then might it not have had hair? Some sort of insulation would have been very useful for surviving the last part of the Permian, when there are signs that glaciers ground down from the pole nearly to the equator.

The adaptations of the later therapsids to running also show a high metabolic rate. Therapsids were probably not warm-blooded, in the sense of having a highly regulated temperature control system, until well into the Triassic. Before that, however, the rate of metabolism might have been high enough to produce body temperatures higher than their surroundings. Many large cold-blooded animals of today can do this. Tuna are an excellent example. The brain, many muscles, and parts of the digestive system of these active fish are maintained at temperatures well above that of the ambient seawater. The rather meaty texture and flavor of tuna is a result of the accumulation of high levels of myoglobin in the fish's muscles, a reflection of their high metabolic rate. This is why fish haters like the author have no problem with tuna, swordfish, and other near-warm-blooded fish.

Sometime during the Triassic, the boundary was crossed between therapsids and mammals. It probably happened more than once, giving rise to the three major mammalian divisions: the egg-laying monotremes, the marsupials, and the placentals. The boundary is usually considered the point at which the several bones of the lower jaw were reduced to one (the others have taken up residence in our middle ears, where they conduct sound to the cochlea). But this boundary is quite elastic and each group may have crossed it more than once. There was such a large advantage to having a single strong jawbone, freeing up the other bones of the jaw to take up different functions, that it apparently occurred several times. Transitional forms have even been found. In the late therapsid *Diarthrognathus* the two different jaw articulations existed side by side.

By the time of the mammal-like therapsids, the world was changing rapidly. These animals must have met problems very similar to those that faced the mammals of South America much later, and they responded to them in ways available to animals with a mammal-like organization and development. Most of the bones of advanced therapsids have their counterpart in the mammalian skeleton. Only a few have altered their functions, though many have altered repeatedly in size and shape. The number, size, function, and position of teeth could and did change frequently. Different parts of the body grew or shrank in size.

As a result of this long evolutionary struggle, it may be that certain evolutionary directions came to be built into our genes, perhaps as long ago as the therapsids, so that parallel evolution can be explained in part as the repeated realization of these ancient lines of increased probability. Had therapsids rather than mammals been set free on the South American island, it is quite possible—even likely—that they would have adapted to that environment in ways very similar to the ways the mammals actually did. Given the time, they might have produced saber-toothed carnivores, single-toed grazers, and so on. They would certainly also, like the mammals, have had the capability of producing very different creatures, as different as the anteaters and the sloths are from the main lineages of mammalian development. As we explore the architecture of our genes, we may find clues to where the well-worn developmental paths lie, how old they are, and how it may be possible for organisms to branch off occasionally in new evolutionary directions.

THE SMOKING GUN

Most of the mutations we know about and that geneticists work with every day are potential-realizing mutations. They alter the highly organized genome in small ways, usually harmful but sometimes beneficial. But new advances in molecular biology will soon allow us to investigate potential-altering mutations as well. We will be able to catch the facilitation of evolution in the act—to find the smoking gun.

Let me mention just one possible line of attack. It involves the Bithorax complex (BX-C), that collection of highly organized genes that regulate the development of the segments of the posterior two-thirds of the fruit fly.

The BX-C shows rather little variation from one member of the species *Drosophila melanogaster* to another. This stands in striking contrast to the substantial variation present in the mimicry complexes of butterflies and to the essentially infinite variation that can be generated from the antibody complexes of mammals. This may be because BX-C is extremely important in development, so much so that mutational change is rare. As a consequence of this genetic uniformity, it has not been possible to connect naturally occurring variations in the Bithorax complex with variations in the anatomy of wild-type flies.

Welcome Bender and his colleagues have looked in detail at the BX-C from two inbred strains of *Drosophila melanogaster*, Canton-S and Oregon-R. The flies are remote descendants of wild flies trapped decades ago in Canton (Ohio) and Oregon, respectively.

The maps they have obtained are not yet complete sequences of the DNA. The sequencing of three hundred thousand bases is still too daunting a task for anyone to undertake lightly, although with newly automated techniques it may soon take only a month or two. Instead, they have constructed a detailed map using restriction enzymes. These enzymes, some of the most powerful tools of molecular biology, cut DNA only in places where a specific sequence of bases is found. Many of these enzymes are known: each is specialized for a different short sequence, and each will cut the DNA wherever that sequence appears.*

Such a map surveys in detail only a tiny fraction of the DNA of BX-C, namely, the short lengths that are recognized by the various restriction enzymes used to construct the map. Nonetheless, it is possible to see whether the maps are of different lengths, whether the restriction sites have appeared or disappeared, and whether the sites have shifted relative to each other.

Bender and his group found that the differences between the two strains were very slight and did not involve any changes in length or shufflings of the order of the restriction sites. The differences could all be explained if one assumed a scattering of single-base substitutions in the DNA of the two strains of the kind that have accumulated between any

---

* Bacteria make these remarkable enzymes, which are called *restriction enzymes* because they restrict or limit the damage caused by bacteriophages. The bacteria use these enzymes to cleave the invading bacteriophage DNA. They make up a kind of primitive bacterial immune system, which molecular biologists have turned to their own remarkable uses. One of these uses is to trace tiny samples of DNA to their owner, who possesses a unique restriction pattern. This technique has begun to revolutionize crime detection. It depends for its success on the immense number of differences among the DNAs of humans.

two members of a species. There was no sign, for example, that transposable elements had inserted or deleted themselves in the course of the time since the two lines had become separated. There were also no indications of inversions.

Without a complete sequence of the complexes from the two lines of flies, it is of course not possible to rule out very small changes of this type. You will recall from chapter 9 the story of the snapdragon *Antirrhinum*, in which stable genetic variants were produced by the insertion and then the imperfect removal of a mobile genetic element. Bits of the repeats at the ends of the element were left behind, like fingerprints at the scene of a crime, and this often had a very large effect on the color of the flower. It will be fascinating to see whether such traces can be found when the BX-C regions from different flies are compared in detail. We find, however, that unlike the flowers of *Antirrhinum*, flies of these two strains of Drosophila look exactly the same even to the tutored eye. There is thus no obvious physical reason to expect any differences between their BX-C sequences. This complex may, as I suggested earlier, be so important that it is highly conserved within a species and shows almost no variation.

Even the absence of variation may tell us something, just as the fact that the dog did not bark in the night told Sherlock Holmes the identity of the murderer. If BX-C is extremely important to Drosophila development, and extremely resistant to evolutionary change, it should be highly conserved among species of fruit fly.

But there may be more substantial differences when we examine the BX-C from more distantly related insects. While the fruit fly's near relatives all have two wings, their more distant relatives—and indeed the majority of flying insects—have four. How does their BX-C differ from that of Drosophila?

Consider the silkworm moth, *Bombyx mori*. The larvae of this magnificent insect undergo a longer and more complex development than those of Drosophila. They must crawl about on the surface of mulberry leaves and must often cling tightly to their edges as the leaves are tossed in the wind. As a result, they have two types of legs through most of this development. There are stumpy legs on each of the three segments that are slated to become the thorax of the adult. Parts of these will play a role in metamorphosis and contribute to the legs of the adult. In addition, four of the abdominal segments have a pair of more fleshy but equally stumpy legs, structures that will disappear completely in the adult.

Several different mutations have been found that lie very close to each other on one of the *Bombyx mori* chromosomes. These affect the number and position of the fleshy abdominal legs and other features of the segments such as the pigment pattern. They are called *E* (*extra legs*) mutants. One of the most interesting of these mutants causes the larva to die before it pupates—but not before it is possible to see that the larva has developed thoracic-type legs on most of its abdominal segments as well. (See figure 11.3.)

This is remarkably similar to what happens when BX-C is deleted in Drosophila. There, you will recall, all the segments posterior to the middle thoracic segment are converted to copies of that segment. Almost certainly, the E-complex of *Bombyx mori* is the homologue of the BX-C of Drosophila.

It will be easy to discover whether this is so, since the BX-C and the genes of the E-complex should be sufficiently closely related that E-complex genes can be fished out of the *Bombyx* DNA using BX-C clones as bait. Some work along these lines has already been done with the flour beetle *Tribolium*. This beetle, which lives well in refined flour, is an economic pest as well as a much-studied laboratory organism. Richard Beeman, working for the U.S. Department of Agriculture, has found through genetic studies that homeotic mutants of both the Bithorax and the Antennapedia type map to the same rather large region on a Tribolium chromosome. Something has certainly happened to the position of these genes in the course of their evolution, and perhaps to other

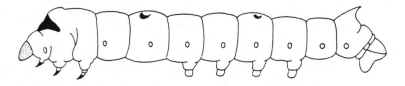

Wild-type Larva

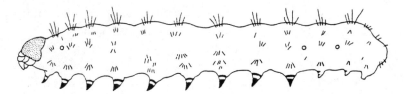

*E*ᴺ/*E*ᴺ **Homozygote Embryo**

Figure 11.3 A wild-type silkworm (*Bombyx mori*) larva and a mutant *Extra legs new additional crescent* (E^N). The mutant dies as a late embryo (as does the Drosophila mutant affected in the same way), but not before it can be seen that most of the segments of the abdomen have been transformed into copies of a thoracic segment. This and other mutants at this locus make it quite certain that the E locus is homologous to the BX-C complex in Drosophila. (Redrawn from figure 8.2 of R. A. Raff and T. C. Kaufman, *Embryos, genes and evolution* [New York: Macmillan, 1983].) Reprinted with permission of Macmillan Publishing Company. Copyright © 1983 by Macmillan Publishing Company.

properties of the genes as well. One intriguing possibility, yet to be proved or disproved, is that Tribolium has only one gene complex of this type, and that this has become duplicated and diverged in the course of the evolutionary divergence of Tribolium and Drosophila. Molecular probes are now being used to investigate this system further.

Differences that are found between these complexes will be prime candidates for potential-altering mutations that have occurred in the course of the separate evolution of these insects. It will be fascinating to see what kinds of differences these are. Perhaps they will explain why *Bombyx* has four wings instead of two; and they may cast light on the origin of the extra abdominal legs of *Bombyx* larvae. The nature of the rearrangements should also give us clues to how they were generated. Will there be traces of the activities of jumping genes? Are there genes or other pieces of DNA present in one sequence that are missing in the other?

If the hypothesis set out in this book is correct, I would expect to find the differences concentrated at particular points that are critical to the specification of certain characteristics of the individual compartments. Further, these are the regions where one might most likely find the traces of mobile elements. It is these points, presumably, at which potential-altering mutations were most likely to have occurred in the course of the separate evolution of the flies and the moths.

These cross-species and cross-group comparisons are some of the most interesting evolutionary studies that will be embarked on in the near future. The kinds of information that can be obtained by comparing developmental genes will be very different from that obtained by comparing protein sequences. This is because developmental genes are much more than just protein-coding regions of the DNA. Their context and arrangement is essential to their function. In the process of understanding them, we have a good chance of finding the smoking gun of evolutionary facilitation.

MIMICS OF SUPERGENES

Mutations of the BX-C type can also be used in a second kind of search for the smoking gun, this time a hunt for potential-realizing mutations. Now that biologists have become so sophisticated in molecular research techniques, I would like to suggest that it is time to go back and repeat some very old experiments. We can get much more information out of them now.

So far as BX-C itself is concerned, there does not appear to be much genetic variation in Drosophila. But genetic variation for the characters controlled by BX-C *does* exist. We simply have to look elsewhere in the genome for it. A series of experiments performed nearly forty years ago, long before the advent of molecular biology, illustrates this point.

C. H. Waddington, a distinguished British embryologist with a highly original mind, carried out the work. He began with the observation that the developmental process is a highly complex one with many steps. At each stage, the genes and the organism's environment interact. If everything goes smoothly, most organisms will develop normally, resulting in the wild type. But if there is something a little bit unusual about the genes *or* if there is a disturbance of the environment or both, then the final phenotype of the organism might be quite different.

An environmental shock, Waddington reasoned, would have the effect of revealing hidden genetic differences among organisms. In other words, differences that were there all the time would only be revealed when the flies were subjected to a shock great enough to disturb the development of the members of the population that carried the appropriate variants.

He set out to test this possibility. Freshly laid eggs of wild-type fruit flies, from a stock containing a fair amount of natural genetic variability, were exposed to ether vapor for a few minutes. While most of the flies developing from these eggs appeared perfectly normal, a few had abnormalities that looked exactly like some of the mutants at BX-C: they showed misshapen and sometimes partially duplicated thoraxes. When the progeny of these abnormal flies were raised without ether shock, they were perfectly normal in appearance. The effect appeared to be a transient one, like the congenital defects seen in some human infants after exposure to drugs or other prenatal hazards.

But Waddington went a step further. He then took other eggs of the abnormal flies and gave them a further ether shock. He found that of the adults developing from these eggs, a larger proportion showed BX-C-like characters than did the unselected flies of the first generation.

Continued selection, along with ether shock each generation, eventually resulted in lines in which all the flies had duplicated thoraxes. These lines continued to show the abnormality even when the ether shock was discontinued. When these strains were analyzed genetically, it was found that the Bithorax-like characteristics were caused by many different genes scattered throughout the chromosomes and not by mutant alleles of BX-C.

How could this have happened? It certainly sounds like the inheritance of acquired characters that we talked about in chapters 3 and 9. Indeed, Waddington had initially suggested such an explanation, but was forced by the outraged screams of his contemporaries to backtrack quickly. The inheritance of acquired characters is an evolutionary no-no. There are, however, two other explanations, one genetically respectable and the other less so.

The first, settled on by Waddington, is based on known genetic principles. Waddington assumed that, scattered through his starting population, there were a number of rare alleles of various genes involved in the development of the thorax. Each was individually rare, and only a few flies had more than one or two of them. Even these flies would develop normally and appear phenotypically normal as adults— *unless* they received an ether vapor shock early in development.

When Waddington selected the progeny that showed the greatest effect of the ether shock, he was actually picking out the flies that carried the largest number of these alleles at various loci, concentrating them as it were. Without the ether shock, he would never have been able to distinguish them from flies that carried few or none of the alleles. Subsequent generations of ether shock and selection concentrated these alleles even further in the selected flies. Eventually, the flies in his selected lines carried so many of these various alleles that a threshold was reached and they developed abnormally even without the ether shock.

Waddington found that the nature of the environmental shock had a great influence on the outcome of the experiment. When he gave eggs a temperature shock rather than an ether shock, he found that some of the progeny lacked a particular vein on their wings. The results of subsequent selection experiments revealed a genetic story very like that for the Bithorax-like condition, except that he ended up with flies stable for missing wing veins rather than doubled thoraxes.

The power of molecular biology can now be used to unravel this old story. We can ask a number of questions. Perhaps the most daring is whether Waddington's genetic-variability explanation was correct. If it was, then inbred strains of flies without much genetic variability should not show the Waddington effect. There is some recent evidence that inbred strains *do* show the Waddington effect, but unfortunately it was not determined just how much genetic variability had been lost through the inbreeding. If these strains really were genetically uniform, then this suggests that at least some of the variation that produced the Waddington effect might have been generated *by the ether shock itself.*

This would have been a very heterodox idea a few years ago. But we have seen again and again that mutations can be produced by much milder environmental conditions than the horrid X-rays and chemicals habitually used by geneticists. If some jumping genes change their rate of jumping a hundredfold as a result of a small temperature change, then why should others not be stimulated to jump by an ether shock?

Please note that this is not the inheritance of acquired characters sneaking in again like King Charles's head into Mr. Dick's memorial. Instead, I am merely suggesting that a sufficiently strenuous environmental shock might trigger the movement of transposable elements, in turn resulting in an increase in a particular class of mutations. Unlike the inheritance of acquired characters, there need be no obvious connection between the nature of the environmental shock and the mutations that might result—though we saw in chapter 9 that there sometimes appears to be such a connection. I cannot predict whether ether shock and heat shock might actually be responsible for some of the variation that Waddington selected. But if they turn out to be a cause, then it will presumably be because intrinsic mutational agents have been disturbed or activated by these environmental shocks. Ether shock would trigger one class of changes, leading to bithorax-like alterations. Heat shock would trigger another class, leading to crossveinless-like alterations.

The tools are at hand to begin such an investigation. We can now use specific probes to ask what happens to copia, P, and other mobile elements of a fly during such an environmental insult. We saw in chapter 9 that these transposable elements are moving all the time in nature, though it has so far proved impossible to make them move except through specific kinds of crosses. Can we stimulate their movement by using environmental shocks like Waddington's?

We can also ask whether different kinds of environmental insults cause different categories of jumping genes to leap about in different ways. Do they, for example, tend to move to regions of the genome that control the Bithorax-like condition when triggered by ether vapor, and to other regions that control wing veins when triggered by heat shock? We should be able to find this out because, once a known mobile element inserts itself into a gene, the gene can be cloned and transformed into the fly in many copies. Effects on the phenotype can then be seen directly.

The experiments suggested here are not only practicable, but many of them are already being done. There are lots of reasons for doing such experiments, since they will cast light on development and on the process of mutation as well as on the mechanisms of evolution. Each supergene will have its own story of evolutionary facilitation and evolutionary repression to tell. As the information begins to come in, the smell of cordite will grow stronger and lead us eventually to the elusive smoking gun.

THE GREAT CHALLENGE

One hundred and fifty years have passed since Darwin set out on his voyage around the world. We have learned a great deal about evolution in that time, but it is nothing compared with what we will learn during the next hundred and fifty. It is quite certain that in that time we will have found the answers to the many questions raised in this book. And in the process we will have learned in great detail how evolution actually happens.

The power this will give us beggars the imagination. It is like the power of the electronics and information revolution since it will enable us to do things that at the moment can be imagined only by science fiction writers. We might, for example, be able to trace the genes back in time and resurrect organisms that have long since become extinct. This has already been done to some extent, not by sophisticated technology but through the millennia-old process of artificial selection.

The ancestor of European cattle was the urus or aurochs, herds of which survived the ice ages in meadows that broke the otherwise trackless wilderness of the Hercynian forest of northern Europe. The immense bulls of this animal were described by Caesar in his *Commentaries* as being nearly as large as an elephant and correspondingly fierce. Whether this was exaggerated or not, the possibility of meeting an enraged aurochs must have made primitive hunters more than usually cautious as they tiptoed through the forest. These immense tracts of ancient trees lasted down to medieval times and live on in our imaginations as the abode of the aurochs, bear, and lion (real) and the troll, unicorn, and hippogriff (presumably imaginary).

The aurochs, too, lasted down to and even through medieval times, with the last certain sightings in the early seventeenth century. It figured in cave paintings and appeared in the bestiary of Konrad Gesner, so we have quite a good idea of how it looked (see figure 11.4). It even appeared in an oil painting by an anonymous Polish artist. But then it vanished from history.

Between the wars, the Berlin and Munich zoos were directed by the brothers Lutz and Heinrich Heck, who performed a pair of fascinating experiments in animal breeding. The results of the experiments were seized upon by the Nazi propaganda mill, but they remain valid nonetheless.

Lutz started with Spanish and French lines of cattle from which fighting bulls were bred; they showed horn conformations similar to that of the aurochs. Heinrich started with Hungarian cattle that were very large and had a build similar to that of the aurochs. Their horns, however, were very different.

In both programs, animals with the most aurochslike characteristics were selected for. Within a few generations, both

Figure 11.4 Konrad Gesner's woodcut of an aurochs. It looks much more mild-mannered than the way Caesar described it. (From *Curious woodcuts of fanciful and real beasts* [New York: Dover Publications, 1971], p. 27).

brothers ended up with aurochslike animals that were indistinguishable from each other and from the pictures of their ancient ancestors. They were not as large or as fierce as Caesar had described them, but they were quite a fair simulacrum of a creature that had disappeared hundreds of years before.

Why was this selection so successful and rapid? It is because the genetic potential for such selection is still present in domesticated cattle in spite of millennia of breeding for other characters. Allelic forms of genes that conferred aurochslike characteristics were still present and ready to be selected for and rearranged in the old combinations. It is hard to rid even highly selected domesticated animals of these ancient alleles. Perhaps they are not all present as such, but instead can easily be resurrected by recombination or by the action of some of the processes, such as jumping genes, that we have talked about in this book.

Is the product of this selection really an aurochs? No, but it is as close as we can come at the present time.

We now know much more than the brothers Heck did. Might it be possible eventually to resurrect the dinosaurs? By all means. All kinds of clues to the dinosaurs remain embedded in the genes of their relatives the birds. It will of course be a good deal more complicated than producing an aurochs. Dinosaurs have left no direct descendants. Their probable genes, including the genes that controlled their development, will have to be pieced together from collateral lines of descent among the birds and the reptiles. We will have to fill in things by educated guesswork, but the existence of evolutionary toolboxes will make the process easier. As a result, it will be quite possible to do and—like so much in science—marvelous fun.

We have only begun to glimpse some of the possibilities inherent in the evolutionary modification of other organisms. Selective breeding has already produced animals and plants without which our civilization could not survive. As a result of the marriage of selective breeding and technology, we are actually entering an era in which many parts of the world are producing too much food, even for our burgeoning population. Unless we damage our environment irreparably, this trend is likely to continue, with the breeding of plants and animals that can outproduce present-day ones by orders of magnitude. And why rely on animals at all for food? If, like me, you enjoy eating great chunks of bloody meat but loathe having to send poor dumb animals to a cruel and unnecessary fate, the answer is simple and soon to come off the drawing boards: the beefsteak bush. The ability to produce muscle and fat tissue does not have to be confined to animals.

There is, in short, nothing to prevent us from designing a

world in which we can live without unnecessary cruelty and needless destruction of the environment. Nothing, that is, except our own ignorance, prejudice, and overwhelming drive to overreproduce ourselves. Because of these failings, we will drive millions of species of animals and plants to extinction over the next few decades. But, if we come to our senses in time and realize the damage we are doing, it may be possible to undo some of it. With care and with growing knowledge, we should be able to reach back into the past and restore much of that diversity of animal and plant life— or at least a perfectly serviceable simulacrum of that diversity—that we are now needlessly squandering.

And what of ourselves?

We can already begin to see how our evolutionary history has shaped our genes. Consider our brief life-spans, which have still not greatly exceeded the biblical threescore and ten. It seems ridiculous that a Volvo, given some care and attention, can outlast most of us. Yet in the course of our evolution we have been selected, even quite recently, for increased longevity.

Our close relatives the chimpanzees rarely live beyond the age of forty. We shared a common ancestor with the chimpanzees between 5 and 7 million years ago, and that ancestor probably had a shorter life-span than either ourselves or the chimpanzees. How did this doubling, or perhaps more than doubling, happen so quickly?

One possibility is that our increased longevity is an accidental result of selection for a slower rate of maturation, which then had a kind of ripple effect on the rest of our life-span. We do not know whether this selection for a slowed maturation rate accounts for all the slowed rate of aging, or whether there was some direct selection for an increased life-span. If the latter, then it may be possible to track down the genes responsible. If there are potential-altering mutations that have led to a doubling of our life-span, it is not beyond the bounds of possibility to suppose that similar mutations could double it again.

Regardless of whether we, or our society, are ready for answers to these questions, they will soon be forthcoming. The pace of science is now such that the answers are arriving even before we have formulated the questions.

The two greatest unanswered mysteries about evolution are the origin of life and the origin of human intelligence. It is proving extraordinarily difficult to make headway on the first problem. But the second problem will soon be answered, because the genes responsible for the evolution of our intellectual development are a part of all of us and can be investigated in exquisite detail.

Our nearest relatives, the chimpanzees, gorillas, and orangutans, carry important parts of the puzzle. Their intellects have continued to develop in ways different from our own since the time of our common ancestors. It will soon be possible to isolate the genes that are responsible for this development and to discover precisely how we differ from the great apes. And perhaps to discover in the process how the toolboxes of genes conferring mathematical ability differ from those conferring musical, artistic, or verbal ability.

Without a doubt, the rapid evolution of our intellects has been facilitated by the ways in which the genes controlling development of our brains are organized. And, as with everything else in evolution, increases in intellectual ability have happened more than once.

Pelycosaurs were undoubtedly more stupid than therapsids, which were undoubtedly more stupid than mammals. Marsupials certainly appear to be less intelligent than most placental mammals of the same size and general habits. The company of a koala bear, endearing though it is, quickly palls. It is probable that all these quantum increases in intellect paved the way for our own immensely rapid intellectual development.

The traces of this facilitation of the evolution of intelligence may still be detectable in the genes. Understanding them may point the way toward our own next intellectual advance, which we will surely take just as soon as we understand how to accomplish it. Whether we are successful or not will depend on how much we understand about evolutionary potential and how to alter it.

These and other discoveries lie in the future. All of them will help us to understand the individual threads of the marvelous tapestry that makes up the evolutionary process. The picture on the tapestry is already visible in outline. How will it look when it is completed? And how will we use the picture that we see?

References

Preface

PAGE

xii   Several of the ideas discussed in this book have appeared in the literature. John H. Campbell has talked about several of the classes of genes discussed here and has even suggested that there may be a category of genes that can "anticipate evolutionary changes," though he gives no examples. John H. Campbell, Autonomy in evolution, in *Perspectives in evolution*, ed. R. Milkman (New York: Sinauer, 1982), pp. 190–201. Stephen Jay Gould and Elizabeth Vrba have resurrected the old idea of *preadaptation*, that characters may appear by chance or for some very different reason, and then be available to take up new functions. See their discussion in S. J. Gould and E. S. Vrba, Exaption—a missing term in the science of form, *Paleobiol.* 8 (1982): 4–15. As this book was wending its way toward the press, Richard Dawkins published a paper in which starting from a very different set of observations, he converged on a number of the ideas set forth here. Richard Dawkins, The evolution of evolvability, in *Artificial life*, ed. C. Langton (New York: Addison Wesley, 1988), pp. 201–20.

xiii   For a discussion of Steve Howell's work on plants that can glow in the dark, see D. W. Ow, K. V. Wood, M. DeLuca, J. R. DeWet, D. R. Helinski, and S. H. Howell, Transient and stable expression of the firefly luciferase gene in plant cells and transgenic plants, *Science* 234 (1986): 856–59.

xiii   How far we have progressed in gene therapy can be glimpsed in some remarkable work from Leroy Hood's lab, in which a gene that corrects a neurological defect has been inserted by microinjection into fertilized mouse eggs. See C. Redhead, B. Popko, N. Takahashi, H. D. Shine, R. A. Saavedra, R. L. Sidman, and L. Hood, Expression of a myelin basic protein gene in transgenic shiverer mice: Correction of the dysmyelinating phenotype, *Cell* 48 (1987): 703–12.

INTRODUCTION

PAGE

3–5   Details about Walter Cannon's life and work can be found in Walter B. Cannon, *The way of an investigator: A scientist's experience in medical research* (New York: Hafner, 1945); idem, *The wisdom of the body* (New York: Norton, 1939).

10   Richard Dawkins has discussed at length the power of natural selection in sorting out sense from random noise in R. Dawkins. *The blind watchmaker* (New York: Norton, 1986).

CHAPTER 1   SOME EVOLUTIONARY MYSTERIES

PAGE

11–16   A selection of the fascinating work on Hawaiian Drosophila is given by the papers in a recent symposium, Diversity in Hawaiian drosophilids: A tribute to Dr. Hampton L. Carson upon his retirement, *Behav. Genet.* 17 (1987): 537–615. A detailed discussion of the earlier work is found in H. L. Carson, D. E. Hardy, H. T. Spieth, and W. S. Stone, The evolutionary biology of the Hawaiian Drosophilidae, in *Essays in evolution and genetics in honor of Theodosius Dobzhansky*, ed. M. K. Hecht and W. C. Steere. (New York: Appleton-Century-Crofts) pp. 437–544.

19   Julian Huxley summarized the neo-Darwinian synthesis in J. Huxley, *Evolution: The modern synthesis* (New York: Harper, 1942).

20–21   A classic paper of Oscar Miller's in which he shows some of the first pictures of genes making RNA (catching them, as he points out, *in flagrante transcripto)* is O. L. Miller, Jr., and B. L. Beattie, Portrait of a gene, *J. Cell Physiol.* 74 (Supp. 1)(1969): 225–32.

25–30   A general survey of malaria and its effect on man can be found in L. J. Bruce-Chwatt, *Essential malariology*, 2d ed. (New York: John Wiley and Sons, 1985); Some glimpses of the current work on malaria parasites and their interaction with the human population are presented in L. H. Miller, J. D. Haynes, F. M. McAuliffe, T. Shirioshi, J. R. Durocher, and M. H. McGinniss, Evidence for differences in erythrocyte surface receptors for the malarial parasites, *Plasmodium falciparum* and *Plasmodium knowlesii, J. Exp. Med.* 146 (1977): 277–81; and T. F. McCutchan, J. B. Dame, L. H. Miller, and J. Barnwell, Evolutionary relatedness of *Plasmodium* species as determined by the structure of DNA, *Science* 225 (1984): 808–11.

31–34   For a general account of the genetically depauperate cheetah population, see S. J. O'Brien, D. E. Wildt, and M. Bush, The cheetah in genetic peril, *Sci. Amer.* 254(5) (1986): 84–92. Some of the most recent work on cheetah populations and their history is described in S. J. O'Brien, D. E. Wildt, M. Bush, T. M. Caro, C. Fitzgibbon, I. Aggunday, and R. E. Leakey, East African cheetahs: Evidence for two population bottlenecks? *Proc. Natl. Acad. Sci. U.S.* 84 (1987): 508–11.

35   Haldane discussed the "nuts and bolts" view of evolution in
            J. B. S. Haldane, A defense of beanbag genetics, *Perspectives in
            Bio. and Med.* 7 (1964): 343–59.
36–44   George Gaylord Simpson has given perhaps the best account
            of the remarkable history of South American mammals in
            G. G. Simpson, Splendid isolation: The curious history of South
            American mammals (New Haven: Yale Univ. Press, 1980).

CHAPTER 2   A NEW WAY OF LOOKING AT EVOLUTION?

PAGE
52–55   The question of species selection and the paleontological view-
            point in evolution is discussed in the following books: S. M.
            Stanley, Macroevolution: Pattern and process (San Francisco:
            Freeman, 1979); idem, *The new evolutionary timetable: Fossils,
            genes, and the origin of species* (New York: Basic Books, 1981).
52   Much has been written about the stop-and-go behavior of the
            evolutionary process. The paper that introduced the term *punc-
            tuated equilibrium* to describe it is N. Eldredge and S. J. Gould,
            Punctuated equilibria: An alternative to phyletic gradualism, in
            Models in paleobiology, ed. T. J. M. Schopf (San Francisco: Free-
            man, 1972), pp. 82–115.
53–55   The knotty problem of sex and why we have it is wrestled with
            at length and inconclusively in J. Maynard Smith, *The evolution
            of sex* (New York: Cambridge Univ. Press, 1978).

CHAPTER 3   HOW NOT TO THINK ABOUT EVOLUTION

PAGE
65–69   The life and thought of Lamarck are described in R. W. Burck-
            hardt, Jr., *The spirit of system: Lamarck and evolutionary biology*
            (Cambridge, Mass.: Harvard Univ. Press, 1977).
69   The pervasive influence of the idea of the Great Chain of Being
            on eighteenth-century thought was first explored in A. O. Love-
            joy, *The great chain of being* (Cambridge, Mass.: Harvard Univ.
            Press, 1942). A more popular survey of this and other ideas
            that preceded Darwin can be found in the anthropologist Loren
            Eiseley's excellent book *Darwin's century: Evolution and the men
            who discovered it* (Garden City: Doubleday, 1958).
70   The growth of the Darwinian world view is brilliantly set forth
            in John C. Green, *The death of Adam: Evolution and its impact on
            western thought* (Ames, Iowa: Iowa State Univ. Press, 1959).
70   The classic work on the devastating impact of Lysenko on So-
            viet science is Zhores A. Medvedev, *The rise and fall of T. D.
            Lysenko* (New York: Columbia Univ. Press, 1969).
70–73   Some recent books on the life and work of Alfred Russel Wal-
            lace (including one—Brackman's—that I think goes too far in
            suggesting that he was badly treated by the Establishment) are
            Harry Clements, *Alfred Russel Wallace: Biologist and social reformer*
            (London: Hutchinson, 1983); Arnold C. Brackman, *A delicate
            arrangement: The strange case of Charles Darwin and Alfred Russel
            Wallace* (New York: Times Books, 1980).

77-81   Books and papers about the Foraminifera are pretty heavy
             going. One slightly more accessible than most is D. G. Jenkins
             and J. W. Murray, eds., *Stratigraphic atlas of fossil Foraminifera*
             (New York: Halsted Press, 1981). A fine example of sped-up
             but still gradual evolution in the Foraminifera is given by B. A.
             Malmgren, W. A. Berggren, and G. P. Lohmann, Species for-
             mation through punctuated gradualism in planktonic Forami-
             nifera, *Science* 225 (1984): 317–19.

86   Problems with fossil dating and interpretation are set forth in
             fascinating detail in Roger Lewin, *Bones of contention* (New York:
             Simon and Schuster, 1987).

91   Sir Richard Burton wrote many fascinating books about his
             adventures. The one that particularly illustrates the points that
             I make here is *The lake regions of Central Africa: A picture of ex-
             ploration* (1860; facsimile ed., New York: Horizon Press, 1961).

93   Did *Australopithecus (Paranthropus) robustus* really use tools?
             Some evidence is discussed in R. L. Susman, Hand of *Paran-
             thropus robustus* from Member 1, Swartkrans: Fossil evidence
             for tool behavior, *Science* 240 (1988): 781–84.

92-93   Many good books have recently summarized human evolution.
             Some recent ones with no particular ax to grind are Bernard
             G. Campbell, *Human evolution: An introduction to man's adapta-
             tions*, 3d ed. (Chicago: Aldine, 1985); Roger Lewin, *Human ev-
             olution: An illustrated introduction* (Oxford: Blackwell, 1984).

98   A fascinating discussion of insect fossils in amber and the sad
             fate of these fossil finds is given in Willy Ley, *Dragons in amber:
             Further adventures of a romantic naturalist* (New York: Viking,
             1951).

CHAPTER 4   THE ROLE OF MUTATION

PAGE
108   Our work on the "yeast to horse" amino acid substitution is
             set forth in C. Wills and H. Jörnvall, Amino acid substitutions
             in two functional mutants of yeast alcohol dehydrogenase, *Na-
             ture* 279 (1979) 734–36.

116-18   The following are discussions of the inconclusive search for
             genetic damage in the offspring of the victims of Hiroshima
             and Nagasaki: W. J. Schull, M. Otake, and J. V. Neel, Genetic
             effects of the atomic bombs—a reappraisal, *Science* 213 (1981):
             1220–27; J. V. Neel, C. Satoh, K. Goriki, J. Asakawa, M. Fujita,
             N. Takahashi, T. Kageoka, and R. Hazama, Search for muta-
             tions altering protein charge and/or function in children of
             atomic bomb survivors: Final report, *Amer. J. Human Genet.* 42
             (1988): 663–76.

119-20   The work of Bruce Wallace and Gerd Bonnier on irradiated
             Drosophila populations can be found in Bruce Wallace, *Basic
             population genetics* (New York: Columbia University Press, 1981).
             Details of the effects on people working in radioactive health
             spas are found in H. Tuschl, H. Altman, R. Kovac, A. Topal-
             oglou, D. Egg, and R. Günther, Effects of low-dose radiation

on repair processes in human lymphocytes, *Rad. Res.* 81 (1980): 1–9.

122–24   The life and work of Barbara McClintock have been set forth for the lay reader by E. F. Keller, *A feeling for the organism: The life and work of Barbara McClintock* (San Francisco: Freeman, 1983).

126   The original papers suggesting the idea of "selfish" DNA are L. E. Orgel and F. H. C. Crick, Selfish DNA—the ultimate parasite, *Nature* 284 (1980): 604–7; W. F. Doolittle and C. Sapienza, Selfish genes, the phenotype paradigm and genome evolution, *Nature* 284 (1980): 601–3.

127   For the original paper on the mutator bacteriophage Mu, see Austin L. Taylor, Bacteriophage-induced mutations in *Escherichia coli, Proc. Natl. Acad. Sci. U.S.* 50 (1963): 1043–50. A fuller discussion of Mu and a good introduction to jumping genes in general are given in James A. Shapiro, ed., Mobile genetic elements (New York: Academic Press, 1983).

132   Some recent papers on the alcohol dehydrogenase system and the way jumping genes affect it are C. E. Paquin and V. M. Williamson, Ty insertions at two loci account for most of the spontaneous Antimycin A resistance mutations during growth at 15 degrees C of *Saccharomyces cereveisiae* strains lacking ADH1, *Mol. Cell Biol.* 6 (1986): 70–79; V. M. Williamson and C. E. Paquin, Homology of *Sacharomyces cerevisiae* ADH4 to an iron-activated alcohol dehydrogenase from *Zymomonas mobilis, Mol. Gen. Genet.* 209 (1987): 374–81.

CHAPTER 5   OF TUXEDOS AND ANTIBODIES

PAGE

137   For one of the best popular introductions to the complexities of the immune system, and a fine piece of very up-to-date science writing, see L. Jaroff, Stop that germ! *Time,* 23 May 1988, pp. 56–64. Some general references include B. Alberts, D. Bray, J. Lewis, M. Raff, K. Roberts, and J. D. Watson, *Molecular biology of the cell* (New York: Garland, 1983), chap. 17; John J. Marchalonis, *Immunity in evolution* (Cambridge, Mass.: Harvard Univ. Press, 1977); F. MacFarlane Burnet and D. O. White, *Natural history of infectious disease,* 4th ed. (London: Cambridge Univ. Press, 1972).

138   A horrifying look at the filth and squalor of Victorian London is found in Henry Mayhew, *London labour and the London poor* (1851; reprint ed., New York: Dover, 1968).

140   Wonderful stories about Metchnikoff and other fighters against disease can be found in one of my favorite books, Paul de Kruif, *Microbe hunters* (New York: Harcourt, Brace, 1926).

CHAPTER 6   AN EVOLUTIONARY TOOLBOX

PAGE

159–60   Much has recently been written about dinosaurs. Two good recent discussions are found in R. T. Bakker, *The dinosaur heresies: New theories unlocking the mystery of the dinosaurs and their extinction* (New York: Morrow, 1986); J. N. Wilford, *The riddle of the dinosaur* (New York: Random House, 1987).

161–63   The three-dimensional structures of yeast and horse liver alcohol dehydrogenases and their remarkable resemblance to each other after billions of years of separate evolution is discussed (though rather technically) in H. Jörnvall, Differences between alcohol dehydrogenases: Structural properties and evolutionary aspects, *Eur. J. Biochem.* 72 (1977): 443–52.

173–75   The nucleotide-binding domain and the evolution of "snap-on" parts of genes are discussed in many papers, among them, M. G. Rossmann, A. Liljas, C.-I. Brändén, and L. J. Banaszak, Evolution and structural relationships among dehydrogenases, in *The enzymes*, ed. P. D. Boyer (New York: Academic Press, 1975), pp. 61–102; A. M. Michelson, C. C. F. Blake, S. T. Evans, and S. H. Orkin, Structure of the human phosphoglycerate kinase gene and the intron-mediated evolution and dispersal of the nucleotide-binding domain, *Proc. Natl. Acad. Sci. U.S.* 82 (1985): 6965–69.

177   Walter Gilbert gives his original proposal for exon shuffling in Why genes in pieces? *Nature* 271 (1978): 501.

178   One of the best discussions of the role of introns in evolution is given by Ford Doolittle, in which he raises many of the points I discuss here. See W. F. Doolittle, The origin and function of intervening sequences in DNA: A review, *Amer. Nat.* 130 (1987): 915–28.

CHAPTER 7   OF BUTTERFLIES AND HANDBAGS

PAGE
181   A general introduction to the subject of mimicry is W. Wickler, *Mimicry in plants and animals* (New York: McGraw Hill, 1968).

184–86   The participation of the heliconids of South America in Müllerian rings and the genetics of these complicated species are set forth most completely in a classic work by Philip Sheppard et al., an especially remarkable paper because it was finally brought lovingly to press by his co-workers almost a decade after Sheppard's death: P. M. Sheppard, J. R. G. Turner, K. S. Brown, W. W. Benson, and M. C. Singer, Genetics and the evolution of Müllerian mimicry in *Heliconius* butterflies, *Phil. Trans. R. London B* 308 (1985): 433–610.

188–91   Detailed discussions of butterfly mimicry are found in D. F. Owen, *Camouflage and mimicry* (Chicago: Univ. of Chicago Press, 1980); E. B. Ford, *Ecological genetics*, 4th ed. (London: Chapman and Hall, 1975). An excellent summary of the monarch-viceroy system is given by L. P. Brower, Ecological chemistry, *Sci. Amer.* 220 (2) (1969): 22–29.

193   A good general discussion of the African butterflies and their ecology can be found in D. F. Owen, *Tropical butterflies* (Oxford: Clarendon Press, 1971).

CHAPTER 8   THE EVOLUTIONARY ORIENT EXPRESS

PAGE

212   Discussions of the probable appearance and variety of Edi-
           acaran life are found in Preston Cloud and M. F. Glaessner,
           The Ediacarian period and system: Metazoa inherit the earth,
           Science 217 (1982): 783–92; A. Seilacher, Discussion of Precam-
           brian metazoans, *Phil. Trans. Roy. Soc. B* 311 (1985): 47–48.

216   James Valentine's suggestions about the origin of the explosive
           Cambrian adaptive radiation are to be found in his Biotic di-
           versity and clade diversity, in *Phanerozoic diversity patterns*, ed.
           J. W. Valentine (Princeton: Princeton Univ. Press, 1985), pp.
           419–24.

223–33   The classic papers laying out the techniques for labeling Dro-
           sophila chromosomes and whole Drosophila larvae and for
           walking down the Drosophila chromosomes are M. L. Pardue
           and J. Gall, Molecular hybridization of radioactive DNA to DNA
           of cytological preparations, *Proc. Natl. Acad. Sci. U.S.* 64 (1969):
           600–604; M. E. Akam, The location of Ultrabithorax transcripts
           in Drosophila tissue sections, *EMBO Journal* 11 (1983): 2075–84;
           W. Bender, P. Spierer, and D. S. Hogness, Chromosome walk-
           ing and jumping to isolate DNA from the *ace* and *rosy* loci and
           the Bithorax complex in *Drosophila melanogaster, J. Mol. Biol.* 168
           (1983): 17–33. A legion of papers using these techniques has
           followed. Discussions of BX-C and ANT-C tend to be rather
           technical, to say the least. The following papers will give brave
           souls an introduction: W. J. Gehring, Homeo boxes in the study
           of development, *Science* 236 (1987): 1245–52; Ginés Morata, E.
           Sánchez-Herrero, and J. Casanova, The Bithorax complex of
           Drosophila—an overview, *Cell Differentiation* 18 (1986): 67–78.

232–37   The papers I have talked about in this section include: L. E.
           Frischer, F. S. Hagen, and R. L. Garber, An inversion that
           disrupts the Antennapedia gene causes abnormal structure and
           localization of RNA's, *Cell* 47 (1986): 1017–23; S. Schneuwly,
           R. Klemenz, and W. J. Gehring, Redesigning the body plan of
           Drosophila by ectopic expression of the homeotic gene Anten-
           napedia, *Nature* 325 (1987): 816–18; W. Bender, M. Akam, F.
           Karch, P. A. Beachy, M. Pfeifer, P. Spierer, E. B. Lewis, and
           D. S. Hogness, Molecular genetics of the Bithorax complex in
           Drosophila melanogaster, Science 221 (1983): 23–29; G. Struhl,
           Near-reciprocal phenotypes caused by inactivation or indis-
           criminate expression of the Drosophila segmentation gene *ftz*,
           Nature 318 (1985): 677–80.

CHAPTER 9   THE INCREASING SOPHISTICATION OF
            EVOLUTION

PAGE

241   My authority on the size of the universe is Steven Weinberg,
           The first three minutes: A modern view of the origin of the universe
           (New York: Basic Books, 1977).

244   The "creationist" experiment was carried out by Marshall S.
           Z. Horwitz and L. A. Loeb, Promoters selected from random
           DNA sequences, *Proc. Natl. Acad. Sci. U.S.* 83 (1986): 7405–9.

248–54   The activities of the Tam3 element in generating different pat-
           terns in snapdragon flowers were worked out by E. S. Coen,
           R. Carpenter, and C. Martin, Transposable elements generate
           novel spatial patterns of gene expression in *Antirrhinum majus,*
           Cell 47 (1986): 285–96.

254–59   For a survey of work on genetic transpositions in trypano-
           somes, see P. Borst, Discontinuous transcription and antigenic
           variation in trypanosomes, *Annu. Rev. Biochem.* 55 (1986): 701–
           32.

260–63   The details of the work of John Cairns's group and of Barry
           Hall on the apparent inheritance of acquired characters (which
           I suggest here to be due to the operation of evolutionary tool-
           boxes) can be found in John Cairns, Julie Overbaugh, and Ste-
           phan Miller, The origin of mutants; *Nature* 335 (1988): 142–45;
           Barry G. Hall, Adaptive evolution that requires multiple spon-
           taneous mutations, I, Mutations involving an insertion se-
           quence, *Genetics* 120 (1989): 887–97. The brilliant work that un-
           raveled the activities of the P element in Drosophila was done
           by F. A. Laski, D. C. Rio, and G. M. Rubin, Tissue specificity
           of Drosophila P element transposition is regulated at the level
           of mRNA splicing, *Cell* 44 (1986): 7–19. The discovery of the P
           elements themselves is a fascinating story involving many labs.
           A survey of the history of these events can be found in W. R.
           Engels, On the evolution and population genetics of hybrid
           dysgenesis-causing transposable elements in Drosophila,
           Philos. Trans. Roy. Soc. Lond B 312 (1986): 205–15.

268   The inexplicable movement of *copia* in one line of Drosophila
           when it stayed virtually quiescent in all the other lines of the
           same lab was discovered by C. Biémont, A. Aouar, and C.
           Arnault, Genome reshuffling of the *copia* element in an inbred
           line of *Drosophila melanogaster, Nature* 329 (1987): 742–44.

269–71   Trudy McKay's work on the effect of P element transposition
           on the rate of evolution in Drosophila is detailed in T. F. C.
           McKay, Transposable element-induced response to artificial se-
           lection in *Drosophila melanogaster, Genetics* 111 (1985) 351–74.

CHAPTER 10   POTENTIAL-REALIZING AND
              POTENTIAL-ALTERING MUTATIONS

PAGE
280–81   Conditions just previous to the ice ages are detailed in a num-
           ber of sources. One with good illustrations is L. B. Halstead,
           The evolution of the mammals (Milan: Eurobook Limited, 1978).

286   Exciting new information is slowly forthcoming about the ex-
           tent of variation at the DNA level within populations. One of
           the most thorough studies is M. K. Kreitman, Nucleotide poly-
           morphism at the alcohol dehydrogenase locus in *Drosophila mel-
           anogaster, Nature* 304 (1983): 412–17.

286   R. C. Lewontin has summarized a great deal of information about the variability harbored in the human species, showing that there is far more variation within races than the relatively small amount that can be attributed to differences between races. See R. C. Lewontin, *Human Diversity* (San Francisco: Freeman, 1982).

CHAPTER 11   CLOSING THE CIRCLE

PAGE

291–99   There is a growing literature on therapsids. A useful summary is given by D. R. Parrington, The origins of mammals, *Adv. Sci.* 24 (1967): 165–73. A well-illustrated survey for general readers can be found in J. C. McLaughlin, *Synapsida: A new look into the origin of mammals* (New York: Viking, 1980).

302–3   Little work has been done on the E locus of the silkworm lately, an oversight that I am sure will quickly be corrected. The source for its genetics is Y. Tanaka, Genetics of the silkworm, *Adv. Genet.* 5 (1953): 239–317. Beeman's exciting work on the homeotic gene complexes of *Tribolium* is set forth in R. W. Beeman, A homeotic gene cluster in the red flour beetle, *Nature* 327 (1987): 247–49.

305–7   A survey of the work of C. H. Waddington on the disturbance of development and subsequent selection of developmental mutants in Drosophila is found in C. H. Waddington, Genetic assimilation, *Adv. Genet.* 10 (1961): 257–93.

307–8   In view of a sudden upsurge of interest in stress-triggered mutations, I cannot forbear to mention a small paper in which I explored the idea some years ago: C. Wills, The possibility of stress-triggered evolution, in *Evolutionary dynamics of genetic diversity*, ed. G. S. Mani (Berlin: Springer-Verlag, 1984), pp. 299–312.

Glossary

Adaptive radiation   The process by which an ancestral species or group of species gives rise in the course of evolution to many different species occupying a variety of ecological niches. Of course, this is a thumbnail description of all of evolution, but isolated cases of adaptive radiation taking place on oceanic islands are particularly striking since their results can be seen clearly.

Alcohol dehydrogenase   This enzyme is mentioned in several parts of the book, particularly in chapter 6. It is widely distributed in nature, and it converts alcohols to their corresponding aldehydes and back again. The alcohol dehydrogenase of yeast works best on ethanol, but our own alcohol dehydrogenases can work on a variety of different alcohols, including alcoholic derivatives of steroids. Enzymes capable of carrying out this activity seem to have evolved more than once.

Allele   One form of a particular gene. Most alleles of a given gene that are found in a gene pool differ only slightly from each other, perhaps by only one or two base differences that have appeared through mutation. Some, however, show larger differences. In the course of evolution, alleles increase or decrease in frequency in the gene pool. Sometimes, one allele will replace another.

Alpha helix   Not to be confused with the double helix of DNA, this helix forms spontaneously from certain sequences of amino acids in proteins. Different proteins have different amounts of alpha helix.

Amino acid   Proteins are made up of long, unbranched chains of amino acids, molecules with both basic and acidic properties. Amino acids vary greatly in their chemistry. Many unusual amino acids have been found in specialized proteins, and some are synthesized in plants as natural pesticides. But a particular set of twenty is found in common throughout the living world, and this is the set that makes up the majority of proteins and is coded by the genetic code.

Antennapedia complex   A gene complex on the third chromosome of *Drosophila melanogaster*, not far from Bithorax, and with sufficient structural similarity to it so that they were both almost certainly derived from a common ancestor. It specifies the properties of the middle thoracic segment and can be active in segments both anterior and posterior to it. The Bithorax complex modifies the effects of the Antennapedia complex, particularly in segments posterior to the middle thoracic.

Antibody   One of a class of proteins known more technically as immuno-globulins, glubular proteins that confer immunity. There are five classes of immunoglobulins, and each confers a different kind of humoral immunity (as opposed to cellular immunity, mediated by *T cells*). Because the antibodies can float freely in the circulation, they can confer protection against invading antigens even if there are no cells present. The body makes millions of different types of antibodies from a much more limited number of genes by methods detailed in chapter 5.

Antigen   Any sufficiently large foreign protein or carbohydrate molecule that can elicit an immune response. Usually only specific parts of the antigens, called epitopes, bind to a particular antibody, but antibodies may bind to different epitopes on the same antigen. As a result, a huge network of antibodies and antigens can form, which may immobilize invading organisms and make them more susceptible to digestion by macrophages.

Australopithecus   A genus of at least four species of hominid, of widely varying stature and habits, and with brains not much larger than that of a chimpanzee. Various of these species flourished in southern and eastern Africa between three and about one-half million years ago, with *A. afarensis* being the oldest. All appeared to have an upright posture. *A. robustus*, with its heavy jaw, large grinding molars, and ridged skull, had a chiefly vegetarian diet; *A. africanus* and *A. boisii* were more omnivorous like modern man. *Australopithecus*, by the way, has nothing to do with Australia; it means "southern ape."

B cell   The cell responsible for the production of free-floating humoral antibodies in the immune system. These cells begin their differentiation in the bone marrow, but migrate to many other parts of the body, particularly the lymph nodes.

Bacteriophage   A virus that preys on bacteria; the name means "bacterium eater."

Bases   The bases are part of the building blocks of DNA and RNA, and provide the specificity for the genetic code. In the double-stranded DNA, adenine always pairs with thymine on the opposite strand, and guanine always pairs with cytosine, resulting in accurate replication of the DNA molecule. But the sequence of bases along one strand of the DNA is not constrained by these pairing rules, so that in theory any message coded in the language of these four bases can be written into the DNA strand by the process of evolution.

Batesian mimicry   A type of mimicry in which a tasty or harmless mimic has evolved a warning coloration, pattern, or behavior very close to that shown by a distasteful or harmful model.

Beta pleated sheet   A structure formed in proteins when zig-zag chains of amino acids lock next to adjacent zig-zag chains. The sequences of amino acids that give rise to to these sheets are very different from those that give rise to alpha helix.

Bithorax complex (BX-C)   A gene complex on the third chromosome of *Drosophila melanogaster* that controls the differentiation of segments posterior to the wing-bearing middle thoracic segment.

Bombyx mori   The silkworm moth.

Cambrian   The period from 550 to 500 million years ago during which all the present animal phyla appeared—along with many others that have since become extinct.

Chromosome   Literally a "colored body," so called because microscopists can stain them easily with acid-loving dyes. These little structures in the nucleus of the cell carry the DNA and the other molecules needed to make accurate copies of themselves. Each chromosome can contain thousands or tens of thousands of genes. Chromosomes stretch out or contract at various times in the life of the cell. The giant chromosomes of Drosophila are ordinary chromosomes that have replicated many times during their stretched-out phase.

Convergent evolution   Evolution in which two very different groups of organisms have converged on the same structure, function, or behavior. In the clearest cases of convergent evolution, the organisms in question had a common ancestor so far back in time that it certainly did not have anything resembling the current structures or functions. A classic example of convergent evolution is provided by the wings of insects and the wings of birds. Their common ancestor was probably a little blob of jelly living in the Precambrian seas, and neither it nor its own ancestors ever had wings. But many other situations are less clearcut, and it is sometimes difficult to draw the line between convergent and parallel evolution.

Copia   One of several families of transposable elements found in Drosophila. In the average fly there may be ten or twenty copies of *copia* scattered around its chromosomes. In a given fly, these elements may occasionally move *en masse* to new places and generations later reinsert themselves into the old sites. It is not known what causes these remarkable movements.

Cretaceous   The "age of chalk," the latter half of the Age of Dinosaurs, lasting from 135 to 65 million years ago. Australia-Antarctica separated from the other continents at the beginning of the Cretaceous, and South America was effectively separated from Africa about two thirds of the way through it.

DNA   Deoxyribonucleic acid, a double-stranded, long-chain molecule that is twisted into the famous double helix and that encodes in its sequence of bases the genetic information of almost all organisms (a few use RNA). Because of its structure, DNA is chemically very stable, as befits the ultimate repository of information in the living cell.

Domain   A part of a protein that has a particular function. A domain may carry the catalytic site of an enzyme, or may form an important part of a structural protein.

Drosophila   A genus of two-winged flies, comprising about 2,500 species world-wide. The tiny flies that land on your melon during a picnic are probably Drosophila. *D. melanogaster* is the species most studied by geneticists, and it is a "weed" that has accompanied man virtually everywhere in the world.

E. coli See *Escherichia coli.*

Ediacaran (sometimes written *Ediacarian*) A period, as yet ill-defined, occupying the 100 million years before the start of the Cambrian. Simple multicellular fossils, many of them very different from anything seen later in the fossil record, were plentiful. None, however, had the calcareous skeletons or shells that marked the beginning of the explosion of different multicellular types in the Cambrian.

Endemic species Animal or plant species found in one circumscribed area of the planet and nowhere else.

Eocene "Dawn age." Extending from about 58 to 35 million years ago. All the major groups of mammals were already present by its start, even though less than ten million years had elapsed since the extinction of the dinosaurs.

Escherichia coli A common bacterium living in the guts of a wide variety of vertebrates, including man.

Evolutionary facilitation (Introduced in this book.) The process by which the genes of organisms have been shaped and arranged in the course of evolution, resulting in an increased likelihood that adaptive mutations will arise. This process can occur as a result of ordinary evolutionary forces, and, as I take considerable pains to point out in the course of the book, is not the result of some mystical ability of evolution to "see" into the future.

Exons Not to be confused with the ecosystem–destroying multinational oil conglomerate. These are the segments of a gene that specify parts of a protein, and that are separated by *introns.*

Extrinsic mutational factors This term I use to describe factors, such as chemicals and radiation, acting from outside the organism to produce mutations. While the organisms subjected to these factors may evolve greater resistance to their effects, the factors themselves of course are incapable of evolving.

Gene pool The collection of genes held in common by a population or a species, of which each member possesses a more-or-less randomized sample. The reader may find this definition superfluous since the term seems to have entered the common tongue recently, as in "Her new husband is definitely from the shallow end of the gene pool."

Genetic code The set of sixty-four code words, each three bases long, that can be formed from the four bases found in DNA. This code is almost universal throughout the living world, though a few small variations have been found. Sixty-one of the words code for the twenty amino acids, with between one and six code words for each amino acid. The other three words code for stop signals and indicate that the end of a protein has been reached.

Genome The collection of genes possessed by an organism.

Glycolytic enzymes A series of enzymes involved in the breakdown of sugars. They release the energy stored in the sugar molecule in small, easily utilized packets. This pathway is very old, since it can support life in many organisms even in the absence of oxygen.

Great chain of being The idea, prevalent up to the eighteenth century, that organisms have been immutably fixed in a ladder of complexity at the time of creation, with man at the top.

Heterozygote An organism with one copy of each of two different alleles, one from each parent, at a given genetic locus. It is worth remembering that an organism can be heterozygous at one locus, but homozygous at another.

Homeotic mutants Mutants in which one part of an organism is modified to resemble another. Such mutants are particularly common among the insects because of the ways in which their developmental pathways have evolved.

Homo erectus A very widely distributed hominid, found in places as diverse as East Africa and eastern Asia. It flourished between one and a half and half a million years ago. Early fossils of this species had rather small brains, but towards the end of its career the brain had grown to overlap in size (though not in detailed structure) that of modern man. There is now excellent molecular evidence that the Asiatic branch of this species was not ancestral to ourselves, but it is still highly probable that the African branch was in or near our direct line of ancestry.

Homo habilis The status of this hominid, which lived in East Africa less than two million years ago, is increasingly in doubt. Originally put firmly and contentiously in the genus *Homo* by Louis Leakey, it now transpires that it had short stature and long apelike arms. Further, different specimens of *H. habilis* differ greatly from each other. Some specimens may be Australopithecines and others something even more primitive.

Homozygote An organism with two copies of the same allele, one from each parent, occupying a particular genetic locus.

Hybridization of DNA If DNA is heated or treated with a strong alkali, it will fall apart into its two strands. The sequence of bases that now lie exposed on each strand will bind to the complementary strand, no matter what its source, and will not bind to strands of DNA with other base sequences. So precise is this specificity that a piece of DNA only a couple of dozen bases long will unerringly find its precise mate (if there is one) in a mass of DNA molecules representing the entire human genome. Or the experimenter can manipulate conditions so that the DNA will find strands to which it is only approximately matched. This is very useful in searching for genes in other organisms that are evolutionarily related to a gene that has been cloned.

Hypolimnas dubius One of a complex of tropical African butterflies, remarkable because it is a Batesian mimic of two quite different distasteful models. Its development into one or the other mimic morphology is controlled by alternative alleles at a supergene complex.

Immunoglobulin   See *antibody*.

Individual selection   This is the selection with which Darwin was chiefly concerned. If an advantageous or disadvantageous gene is expressed in an individual, this will immediately affect its likelihood of leaving offspring. This can be contrasted to species selection, in which the eventual day of reckoning may be delayed for hundreds of thousands or millions of years.

Intrinsic mutational factors   I have given this term to factors that live within the cell or that invade it, and that produce mutations, like retroviruses and transposons. These factors are often themselves capable of evolving.

Intron   A segment of a gene which is transcribed into RNA but not translated in turn into protein. Other parts of the RNA, essential to its function, are also not translated into protein; the crucial difference between introns and these regions is that the RNA copies of the introns are specifically removed by special protein-RNA enzymes before the RNA can function. At first blush, the introns appear quite unnecessary to the functioning of the gene, and yet most genes in higher organisms have them. They may (or may not) be involved in the process of building new genes through the process of exon shuffling, or as I prefer to call it, domain shuffling.

IS element   A short piece of DNA, found in bacteria, that can insert copies of itself elsewhere in the bacterium's genome, often producing mutations.

Jumping genes   Genes, or collections of genes, that have the capability of making copies of themselves, copies that are inserted elsewhere in the genome and that often produce dramatic mutations. They are found everywhere in the living world. Various types include retroviruses, transposable elements, transposons, and insertion sequences or IS elements.

Lamarckism   As commonly used, a short and quite inaccurate term for the inheritance of acquired characters.

Locus   This term is a bit confusing to people dipping into the gene pool for the first time. It is simply the location of a given gene on a chromosome. The term dates back to the days when genes were thought to be little balls of protein (or something) strung out along the chromosome like beads on a string. The locus would be the place occupied by one of the little balls. We now know that genes are messages embedded in a long strand of DNA, so the locus is really the part of the DNA in which that particular message is to be found.

Macroevolution   Evolution resulting in large changes, often taking millions of years to accomplish.

Macrophage   A cell capable of engulfing and digesting an invading foreign object. Macrophages interact with T cells.

Marsupial   One of a group of mammals in which most of fetal development takes place in an external pouch. This is because they lack the combination of maternal and fetal tissue that gives rise to a placenta. Some advanced marsupials have a placenta-like structure that helps to nourish the embryo, but it is not elaborate enough to allow development to take place entirely in the womb.

Microevolution   Small evolutionary steps that are the result of the replacement of alleles at various loci by similar alleles, either already present in the population or produced by minor mutational alterations. This is the sort of evolutionary change that can be accomplished by plant and animal breeders, or by workers using experimental organisms in the laboratory.

Mu   This *bacteriophage* can insert copies of itself anywhere in the genome of the bacterium *Escherichia coli*, often producing mutations at a very high rate.

Mutation   Any inherited alteration in the genome of an organism, aside from the mixing of genes produced by the process of genetic recombination. I have deliberately made this definition very broad in order to accommodate the bewildering variety of mutations that the reader is introduced to in chapter four.

Müllerian mimicry   A type of mimicry in which two or more distasteful or dangerous species have converged on the same warning color, pattern, or behavior, gaining mutual benefit. In the tropics, whole clusters of species have converged on the same appearance, forming Müllerian rings. Batesian mimics are often found taking part in such rings, and gain added benefit from the protection afforded by a cluster of Müllerian mimics.

NAD   This is not a misprint for DNA, but stands for nicotinamide adenine dinucleotide, a molecule that is an important helper in many reactions involving oxidation and reduction. Many enzymes bind NAD and use it as a cofactor in the reaction they are catalyzing.

Neanderthal man   These hominids, who flourished in the Middle East and Western Europe between 70,000 and 35,000 years ago, were very similar in appearance and behavior to ourselves, though opinion has lurched back and forth several times since their discovery 130 years ago about how similar to us they really were. Current opinion is that they were distinguishably different in many ways; but different or not, they had an elaborate tool-using culture that may have included ritual burial.

Orthogenesis   The idea that there is a built-in directionality to evolution.

P element   A transposable element, found in natural populations of *Drosophila melanogaster*, that produces a variety of genetic alterations in the progeny if males that carry copies of it are mated to females that do not.

Pangaea   At least twice in the history of the earth all the continents came together to form a supercontinent. Pangaea I's origins are lost in antiquity, but it began to break up at the beginning of the Cambrian. Pangaea II came together at the time of the therapsids some 280 million years ago and began its slow breakup at the start of the Age of Dinosaurs about 180 million years ago.

Papilio dardanus   A butterfly species found through a wide territory in southern and eastern Africa. While the males are similar throughout its range, the females are Batesian mimics of different distasteful models in different regions. The development of these various mimetic forms is controlled by a series of alleles of a supergene complex.

Parallel evolution   Evolution in which two geographically separated groups of organisms have evolved through a series of similar stages, continuing to resemble each other despite their separation. Much must be known about the fossil records of the organisms involved in order for a good case to be made for parallel evolution. The hooves of the horses of North America and the horselike mammals of South America are good examples of parallel evolution, since their ancestors both had feet with a number of toes, all but one of which were lost or much reduced in the course of their adaptation to a grassland environment. But whether their other horselike adaptations evolved in parallel fashion or were convergent instead is much more difficult to assess because of the fragmentary nature of their early fossil record.

Pelycosaurs Reptile ancestors of the therapsids. They flourished at the end of the Carboniferous period some 310 million years ago.

Pleistocene The epoch extending from about two million to ten thousand years ago, marked by numerous glaciations.

Pliocene An epoch with a quite stable and mild climate, extending from twelve to two million years ago, during which mammals reached their greatest diversity.

Potential-altering mutations (A term introduced in this book.) Mutations altering the structure or organization of a genetic toolbox, a reorganization which in turn influences the probability that certain classes of potential-realizing mutations will occur subsequently.

Potential-realizing mutations (A term introduced in this book.) Mutations that arise in a genetic toolbox, and that are more likely to occur because of the structure of the toolbox itself. If the toolbox were arranged in a different fashion, or if the genetic components of the toolbox were disoriented with respect to each other, the likelihood of such mutations would be much less.

Precambrian The time extending from the origin of life up to the appearance of plentiful and diverse multicellular organisms at the start of the Cambrian, some 550 million years ago.

Promoter sequence A DNA sequence, slightly in front of the region coding for the gene itself, which the enzyme RNA polymerase recognizes and to which it attaches in order to begin the production of an RNA message.

Punctuated equilibrium The name given by Niles Eldredge and Stephen Jay Gould to the phenomenon, very widespread in the fossil record, of long periods of stasis in the appearance of organisms, punctuated by brief periods of dramatic change. These latter occur so swiftly that fossils chronicling intermediate types cannot usually be found. The naming of this phenomenon by Eldredge and Gould sparked much argument at three levels. First, how widespread is it? (It appears indeed to be very widespread, though not universal.) Second, are the step-by-step evolutionary processes of beanbag genetics sufficient to explain these sudden changes? (The jury is still out.) Third, what is the cause and what are the genetic consequences of the long periods of stasis? There are probably many causes, but one remarkable and little-investigated phenomenon that has relevance to the arguments in this book is that even after long periods of stasis the genetic variability in most species is undiminished. The preservation of this variability permits evolution to continue.

Quantitative (multigenic) variation Phenotypic variation in which the genetic component of the variability is contributed by alleles at several genes rather than one.

Retrovirus A virus that encodes its genetic information in *RNA*, but that once inside the cell uses the cell's machinery along with its own to make a DNA copy of the RNA. This DNA copy may sometimes become integrated into the cell's own genes, where it may lie undetected for generations. The process of incorporation can produce mutations, as can the subsequent removal of the viral genome. Sometimes the DNA copy of the retrovirus can become permanently inactivated through mutations that it has acquired, and as a result, becomes a permanent fixture of the host's genome. It should be mentioned that only a minority of known retroviruses have been associated with particular diseases.

RNA This single-chained molecule is copied from DNA and translates the information encoded in it into proteins. In some viruses, including the AIDS virus, RNA carries the genetic information. RNA is much less stable than DNA, so this is one reason why these viruses evolve so quickly.

Somatic mutation A mutation which occurs in a body cell rather than a cell in the germ line, and which therefore cannot be passed on to the next generation.

Species selection Selection in which the species is the unit rather than the individual. When a species becomes extinct, it takes with it all the accumulated differences that separate it from other species, just as when an individual dies without offspring, it takes with it all the accumulated mutations, even highly advantageous ones, that it has acquired during its lifetime.

Supergene complex A set of closely linked genes that, because they are near each other on the chromosome, are rarely broken up by recombination and are usually passed on to the next generation as a unit.

T cell A family of cells that have a number of functions. Helper T cells pass increasingly specific information about the nature of an invading antigen to the appropriate B cells and killer T cells, causing them to multiply. Killer T cells can destroy invading cells directly.

Tam3 A transposable element found in a few copies in the genome of the snapdragon *Antirrhinum majus*.

Teleology From the Greek *telos*, meaning end. In evolution, the idea of progression towards a goal.

Therapsids A diverse group of organisms that flourished for 100 million years before the dinosaurs and was eventually displaced by them. Some, though not all therapsid lineages showed mammalian characteristics, and evolutionary links between these therapsid lineages and mammals have been found.

Transposable element This term usually refers to those elements, found in higher organisms, that are rather like retroviruses in their general makeup and in their effect on the genome. The various elements found in Drosophila, such as P and copia, fall into this category. It is possible that, like retroviruses, some of them may be responsible for diseases as well as for genetic damage, but our knowledge of Drosophila diseases is inadequate, to say the least.

Transposon A set of genes found in bacteria and flanked by a pair of IS sequences. Copies of a transposon can be inserted elsewhere in the bacterial genome.

Triassic The first third of the Age of Reptiles, extending from 230 to 180 million years ago. It is notable because it saw the transition of various therapsid lineages into mammalian lineages, and because by the end of it the earliest dinosaurs were beginning to challenge the ascendancy of the therapsids.

X-ray crystallography The technique by which the three-dimensional structure of crystals, including crystallized proteins, can be determined.

Index

Index entries in boldface are defined in the glossary.

Ability to evolve, evolution of, 56–58

Acquired characters: evolution of, 260–63; inheritance of, 68–69, 306, 308

Adaptations: cold-weather, 281, 282–83; convergent evolution of, 42–44

Adaptive explosion, Cambrian, 215–17, 247

Adaptive radiation, 15, **326;** evolutionary toolboxes and, 289–90; of Foraminifera, 79; in Hawaii, 35, 288–90; of mammals after demise of dinosaurs, 38, 291; of therapsids, 294–97

Adenine, 104

ADP, 172

Africa, types of malaria in, 25–29

African sleeping sickness, 255–57

Aggunday, I., 318

AIDS (Acquired Immunodeficiency Syndrome) virus, 155–56, 254

Akam, M.E., 323

Alberts, B., 321

Alcohol, production of, 102–3

Alcohol dehydrogenase, 326; bacterial, 167; conservation of three-dimensional structure in course of evolution, 163; horse vs. yeast, 107–8, 109, 161–63, 165, 166, 167; mutants of, 103–10

Algae, 214

Allele, 24–25, **326;** Duffy-negative, 26–29; genetic recombination and effect of, 34; presence of ancient, in present-day animals, 311–12; spread through population of new, 33–34; supergene, 204–5

Alpha helix, 171, 172, 173, 174*n*, **326**

Altman, H., 320

Amauris butterflies, mimicry by *Hypolimnas dubius* of, 194–200

Amauris echeria, 201

Amauris niavius, 194

Amauris tartarea, 194

Amber, insect fossils in, 98–99

American Civil Liberties Union, 61

Amino acid, 326; alpha-helix, 171, 172, 173, 174*n*, 326; arrangement of, in proteins, 171–73; beta pleated sheet, 171, 172, 173–74, 327; in immunoglobulin molecule, 150; sequences in horse vs. yeast alcohol dehydrogenase, 161–63, 165; specified by genetic code, 105; substitution of, in proteins, 105–10

Amoebas, 219

Amphibians, 292

Anteaters, 38, 42, 292

Antennapedia complex (ANT-C), 227, 229–31, 235, **326–27;** function

Antennapedia complex (ANT-C)
(*continued*)
of, 229; *fushi tarazu* gene, similarities to, 236–37
Antennapedia (mutant), 230–31
Antibody, 137, 139, 144–49, **327;** against trypanosome, 257; assembly of gene for, 150–53; C region, 150; H and L chains, 150; IgG, 143; IgM, 143–44; J, C, and V genes, 150–53, 156–58; number of antibody genes compared to, 148–49; potential-altering mutations in antibody system, 277; production of excess, 247; specificity of, 141–42; V region, 150, 151, 153, 157–58. *See also* Immune system
Antigen, 139, 247, **327;** Duffy, 26; specificity of antibody to, 141–42; T cells and, 140
Antigenic switches in trypanosomes, 257–59
Antirrhinum, 301
Antirrhinum majus, 249–53
Aouar, A., 324
Arch in skull of therapsids, 293
Argon-40, 86
Armadillos, 38, 42
Arnault, C., 324
Arthritis, 142
Arthropods. *See* Insects
Asakawa, J., 320
Asexual reproduction, 54
Atomic Bomb Casualty Commission (ABCC), 116, 117
ATP, 170, 172
Aurochs, 310–12
Australia: aberrant offshoot of mammals in, 292; convergent evolution in, 43–44; marsupials and placentals in, 37–38, 292
Australopithecus, 327
Australopithecus africanus, 92
Australopithecus boisei, 92
Australopithecus robustus, 92–93
Autoimmune response, 142
Automimicry, 191

Back mutations, 249
Bacterial infections, 137–39

Bacterial pneumonia, 154
Bacteriophage, 327; Mu, 126–30, 133, 252, 268, 332; restriction enzymes and, 300*n*
Bakker, Robert T., 160, 321
Balanced Treatment Act (Louisiana), 62
Balanced Treatment of Creation–Science and Evolution–Science Act (1981), 59–60, 61
Banaszak, L.J., 322
Barnwell, J., 174, 318
Bases, 104–6, **327**
Bates, Henry Walter, 70, 186
Batesian mimicry, 208, **327;** dependence on fate of model species in, 186–88; of monarch butterfly by viceroy, 188, 189, 191–93
Bateson, William, 229
Bauriamorphs, 294
B cell, 141–42, 143, **327**
Beachy, P.A., 228, 323
Beagle (ship), 72
Beanbag genetics, 35, 109
Beattie, B.L., 21, 318
Beeman, Richard W., 302, 325
Beerbower, J.R., 76
Behavior, effect of supergene complex, 200
Bender, Welcome, 233, 300, 323
Benign tertian malaria, 25–29
Benson, W.W., 322
Benz, Karl, 274
Berggren, W.A., 320
Beta pleated sheet, 171, **327;** six-stranded, 172, 173–74
Biémont, Christian, 268, 324
Biochemistry as science, beginning of, 169–70
Bithorax complex (BX-C), 230, **328;** absence of variation within species, 299–301; cloning of, 233; E-complex homologous to, 302, 303; *fushi tarazu,* similarities to, 237; genetic variation for characters controlled by, 305–8; gypsy element and, 251; major function of, 227–29; mapping to organism of, 232–34, 235
Bithorax mutants, 233
Bithoraxoid mutants, 233
Blake, C.C.F., 172, 322
Blennies, 183, 184

Blind Watchmaker, The (Dawkins), 10
Bombyx mori, 301–2, 303, 304, **328**
Bonnier, Gerd, 120, 320
Borhyaenids, 40–42
Borst, P., 324
Boyer, P.D., 322
Brackman, Arnold C., 319
Brändén, C.-I., 174, 322
Bray, D., 321
Breeding, selective, 310–13
Brower, Jane van Zandt, 191
Brower, Lincoln P., 189, 191, 322
Brown, K.S., 322
Bruce, David, 256
Bruce-Chwatt, L.J., 318
Bruecke, E. von, 3
Buchner, Eduard, 169–70
Buffon, Comte de, 65
Burckhardt, R.W., Jr., 319
Burnet, F. MacFarlane, 321
Burns, John M., 180
Burton, Richard, 91–92, 320
Bush, M., 318
Butterflies, mimicry in. *See* Mimicry
    in butterflies, evolution of
BX-C. *See* Bithorax complex (BX-C)

Caesar, 310
Cairns, John, 7, 260, 324
Calactin, 190
Calotropin, 190
Cambrian, 328; adaptive explosion
    during, 215–17, 247
Camouflage, 181–83. *See also* Mim-
    icry in butterflies, evolution of
Campbell, Bernard G., 320
Campbell, John H., 7, 317
Cancers of immune system, 148
Cannon, Walter B., 3–6, 318
Canton-S strain of *Drosophila melan-
    ogaster,* 300
Carbon-14 dating method, 86
Carbonic anhydrase, 165–66
Carboniferous, 100, 222, 292
Cardiac glycosides, 190–91
Caro, T.M., 318
Carpenter, R., 250, 324
Carson, Hampton L., 14, 15, 318
Casanova, J., 323
Castellani, Count Aldo, 256
Cattle, ancestor of European, 310

C (constant) immunoglobulin gene,
    150–53, 156, 157, 277
Cheetah, African, loss of genetic
    variability in, 31–34
Chemical mutagenesis, 121–22
Chondrodystrophic dwarf muta-
    tions, 104, 113
Chordates: development of, 218;
    segmentation in, 220–21
Chromosome, 328; giant, of Droso-
    phila, 123, 224–26; homologous,
    146; information carried on, 22;
    walking down the, 233. *See also*
    DNA; RNA
Chromosome doubling in verte-
    brates, mutation causing, 115
Church and state, separation of, 60,
    62
Clarke, Arthur C., 260
Clarke, Cyril, 203–5
Cleaner stations on coral reefs, 183
Clements, Harry, 319
Climate: cold-weather adaptations,
    281, 282–83; fossils and ancient,
    100; ice ages, preparation for,
    280–87
Clones: of BX-C, 233; produced by
    cancers, 148
Closely linked genes, 197–200
Cloud, Preston, 323
Coen, E.S., 250, 324
Colbert, E.H., 293
Cold-weather adaptations, 281,
    282–83
Colorado River, 212–13
Commentaries (Caesar), 310
Common cold virus, 155
Complement system, 141
Computer viruses, 125–26
Condylarths, 38, 42–43
Congenital effects of radiation, 116–
    17
Consanguinity of parents, effect of,
    118
Constant (C) region of immuno-
    globulin molecule, 150
*Contributions to the Theory of Natural
    Selection* (Wallace), 72
Convergent evolution, 42–44, 79,
    328
Copia, 234, 268, 308, **328**
Coral reefs, cleaner stations on, 183
Core wars programs, 125–26

Corn, McClintock's experimental
   work on, 122–24
Creationism, 9–10; argument
   against evolution, 240–42; as sci-
   ence, 59–65; standard evolution-
   ist answer to, 242–43
Creationist experiment, 240–48
Creation Research Society, 60, 61,
   64, 110
Cretaceous, 78, 79, 80–81, **328**
Crick, F.H.C., 321
Cro-Magnon man, 97
Crossing-over, unequal, 146, 147,
   176
Cross-species and cross-group com-
   parisons, 302–4
Cryptic genes, 260–62
Crystallography, X-ray, 164, 335
Cuénot, Lucien, 76
Cugnot, Nicholas-Joseph, 274
Cuvier, Georges, 38–39, 69
Cynodonts, 296
Cytosine, 104

Dachshund, chondrodystrophic
   dwarf mutation in, 104, 113
Dame, J.B., 318
Danaus plexippus (monarch), mim-
   icry of by viceroy butterfly, 188,
   189, 191–93
Darrow, Clarence, 62n
Darwin, Charles, 7, 8, 110, 112,
   309; criticism of Lamarck, 69; on
   Foraminifera, 82–83; misclassifica-
   tion of litopterns by, 42; natural
   selection and, 16–19, 71, 73; *Ori-
   gin of Species, The,* 16, 70, 72n;
   survival of fittest idea, 17
Darwin, Erasmus, 67
Darwinian fitness, 51
Darwinism, social, 17
Darwinism (Wallace), 72
Dating of fossils, methods of, 86–87
Davies, John, 59
Dawkins, Richard, 10, 317, 318
Delbrück, Max, 11
DeLuca, M., 317
Dentition, therapsid, 294–97
De Vries, Hugo, 111–14
DeWet, J.R., 317

Diarthrognathus, 298
Diatoms, 93
Dichotomous key, 65
Dicynodonts, 294–96
Dimetrodon, 293
Dingo, 40
Dinosaurs, 28, 291, 292, 312; characteristics of, 99, 159–60; extinction of, 79, 99–100, 159; fossils,
   97, 159–60;
Dinosaurs, Age of, 78, 79, 99–100
Directionality of evolution, fallacy
   of, 74–84
DNA, 20–23, **328;** bases in, 104–6,
   327; of BX-C, 300–301; chemical
   mutagens acting directly on, 121–
   22; coding for genes for RNA,
   electron micrograph of, 21; deletion of in immunoglobulin gene
   formation, 157; experiment on
   random sequences in, 244–48;
   hybridization, 330; introns, 172,
   177–78, 264, 266, 267; making alterations in, 105–10; measuring
   variation within species in, 285,
   286; mutations caused by Mu,
   127–28; nucleotides in, 22–23;
   parasites of, 7, 133–34; parasitic,
   126; repair systems, 119; restriction enzymes and, 300. *See also*
   Insertion sequence (IS) elements
Doble, Abner, 275
Dobzhansky, Theodosius, 62n
Dollo, Louis, 75–76
Domain: as building blocks in protein evolution, 179; of protein,
   172, 173, **328**
Domain shuffling, 178–79
Domesticated organisms, loss of
   jumping genes in, 268
Doolittle, W.F., 321, 322
Drosophila, 114, 158, **328–29;** enzymes, 166–67; Goldschmidt's
   work on, 114; Hawaiian, 13–16,
   288–90; homeotic mutants in,
   229–32; jumping genes in, 123;
   resistance to radiation, 120; segment duplication in, mutants affecting, 235–37; segment
   evolution in, 237–38
Drosophila cyrtoloma, 14
Drosophila hamifera, 15

Drosophila melanogaster, 13, 14; BX-C from, 227–30, 232–34, 235, 237, 251, 299–301, 302, 303, 305–8, **328;** as developmental tool, 222–26; hybrid dysgenesis from mating of laboratory and wild strains, 264–68; IS elements in genome of, 124

Duckbill platypus, 292

Duffy antigen, 26

Duffy-negative gene, 26–29

Duplication: gene, 145–46, 147, 153, 172, 176; segment, 235–37

Durocher, J.R., 318

Dwarf mutations, chondrodystrophic, 104, 113

East Africa: Burton's accounts of, 91–92; human and hominid diversity in, 91–93; Lake Turkana basin of, 85–90

Edaphosaurus, 293

Ediacara Hills (South Australia), 212

Ediacaran, 215–18, **329**

E (extra legs) mutants of *Bombyx mori*, 302, 303

Egg, D., 320

Einstein, Albert, 9, 70

Eiseley, Loren, 318, 319

Eldredge, Niles, 52, 319

Endemic species, 329

Engels, W.R., 324

England, bacterial infection in during nineteenth century, 137–38

Engrailed, 231

Environment: phenotype and, 23; rate of specific mutations and, 262

Environmental change, 48–49, 50; during insect's development, 231–32; mutations produced by shock of, 53, 305–8; sexual reproduction and dramatic, 55; speciation of freshwater molluscs in response to, 89; species evolution and, 52–53, 84–96; during vertebrate development, 232. *See also* Adaptive radiation

Enzymes: evolutionary relationship of, 165–69; function of, 166; glycolytic, 169–74, 179, 277, 330; restriction, 300

Eocene, 39, **329**

Equal-time laws, 62, 63

Equilibrium, punctuated, 52–53, 333–34

Escherichia coli, 329; bacteriophage Mu infection in, 127; experiment on gene conferring resistance to tetracycline of, 244–47; IS elements in genome of, 124; salicin utilization by mutants of, 260–62

Estuaries, 212–15

Ethiopia, *Papilio dardanus* found in, 202, 203

Evans, S.T., 172, 322

Evolution, 45–101; of ability to evolve, 56–58; as accumulation of changes in gene pool, 34; of acquired characters, 260–63; basics of, 16–25; beanbag view of, 35, 109; convergent, 42–44, 79, 328; creationist experiment and, 240–48; ease of increasing with time, 7, 45–50; explosive, 11–16, 79; fossil record as complete picture of, fallacy of, 96–101; homeotic genes and, 234–38; increasing sophistication of, 239–72; of interaction between mutation-producing factors and genome, 7; Lamarck and, 66–70; lay perception of, 65; misconceptions about, 59–101; movement toward goal in, fallacy of, 74–84; organisms directing their own, fallacy of, 65–73; parallel, 41, 290–91, 298–99, 332–33; public uncertainty about, 9–10; religion and, 9–10, 59–65; repetition of, 36–44; smooth and direct, fallacy of, 84–96; snapdragon transposable elements and, 248–54; species selection, sex and, 53–55; species selection and individual selection in, 50–53; speed of, 55–58; high-energy radiation and, 121; synthesis of genetics and, 18–25; Wallace's ideas of, 70–74. *See also* Darwin, Charles

Evolutionary facilitation, 6–7, 44, 46–50, **329;** antibodies illustrating, 157–58; environmental

Evolutionary facilitation (*continued*)
change and, 48–49, 50; finding
smoking gun of, 299–304; genes
and process of, 6–7; of human
intelligence, 314–15; immune system and, 137–49; in insects of
Hawaii, 288–90; in mammals of
South America, 290–99; mimicry
supergene complex and, 205–9;
at molecular level, 161–79; objections to, 46–47; at organism level,
181–209; parallel evolution and,
291, 298–99; process of, 262–63;
in species selection model, 282

Evolutionary potential, mutations
that change, 276–87

Evolutionary toolboxes, 159–79;
adaptive radiation and, 289–90;
changes in potential for evolution
and, 276–79; degrees of sophistication in, 271; evolution of acquired characters and, 260–63;
modification of genome by, 174–
79; for glycolytic enzymes, 169–
74, 179; protein domains and,
179; of snapdragons, 248–54; of
supergenes BX-C and ANT-C,
230; in trypanosome, affecting
host's immune system, 259

Exons, 177, **329**

Exon shuffling, 178

Exoskeleton, insect, 221

Explosive evolution, 11–16, 79;
Cambrian adaptive explosion,
215–17, 247. *See also* Adaptive radiation

Extinction: of dinosaurs, 79, 99–
100, 159; event at end of Permian, 294; events shaping history
of mammals, 281, 291–92; too little genetic variability and, 31–34

Extrinsic mutational factors, 116–
22, **329;** chemicals, 121–22; radiation, 116–21. *See also* Environmental change

Facilitation, evolutionary. *See* Evolutionary facilitation

Falciparum malaria, 25–27, 29

Feline infectious peritonitis (FIP),
31, 34

Fermentation: cell-free, 169–70;
process of, 102–3

Filter feeders, 214, 218, 219

Fish: camouflage ability in, 181,
182; chromosome doubling in ancient, 115; mimicry of cleaner,
183, 184

Fisher, Ronald A., 34, 54, 188

Fitness, Darwinian, 51

Fittest, survival of the, 17

Fitzgibbon, C., 318

Flounder, camouflage ability of,
181, 182

Foraminifera, 77–84; adaptive radiations of, 79; Darwin on, 82–83;
examples of Cretaceous and
modern, 79, 80–81; fossils, 77–84,
93–97

Ford, E.B., 201, 322

Ford, Henry, 275, 277

Fossil record, 85–101; incompleteness of, 96–101; process of speciation seen in, 84–96

Fossils: ancient climate and, 100;
from ancient estuaries, 212–13;
from Australia, 38; dinosaur, 97,
159–60; directionality of evolution
and, 75–77; Foraminifera, 77–84,
93–97; of Great Rift Valley, 85–
90; of *Gryphaea*, 75–76; hominid,
97–98; insect, 98–99, 222; in
South America, 38–39

412 transposable element, 234

Fragmentation of genes, 149,
153

Frischer, L.E., 323

Fritillaries, 189; viceroy butterfly,
188, 189, 191–93

Frost, Robert, 239

Fruit flies, 13; mutations in, 227–32.
See also Drosophila

Fujita, M., 320

Fundamentalist Christians, 9–10,
59–65. *See also* Creationism

Fushi tarazu gene, 236–37

Gall, Joseph, 226, 323

Galton, Francis, 19

Garber, Richard L., 230–31, 323

Gehring, Walter J., 231, 323

Gemmules, 19

Gene duplication, 145–46, 147, 153, 172, 176
Gene fragmentation, 149, 153
Gene pool, 23–30, **329;** evolution as accumulation of changes in, 34; genetic variability in, 30–35; mutations in, 25–30
Genes: closely linked, 197–200; cryptic, 260–62; different classes 23; evolutionary facilitation process and, 6–7; genealogy of, 165–69; homeotic, evolution and, 234–38; J, C, and V, 150–53, 156–58; Mendel's laws, 18; organization of, 136–37. *See also* Jumping genes
Genetic code, 105, 121, **329**
Genetic recombination, 34, 114, 146, 147
Genetics: beanbag, 35, 109; as experimental science, beginning of, 112; population, 34–35, 285; synthesis of evolution and, 18–25
Genetic uniformity in species, 283–87
Genetic variability: perils of too little, 30–35; Waddington effect and, 307
Genetic variation, P elements and, 270–71
Genome, 23, **329;** interaction of mutation-producing factors and, 7; role of toolboxes in shaping, 174–79
Genotype, 23
Georgiev, G.P., 224
Gerasimova, T.I., 224
Germ-cell damage from radiation, 118–20
Germ line, P elements confined to, 266–68
Gesner, Konrad, 310, 311
Giant chromosomes, 123, 224–26
Gilbert, Walter, 7, 177, 322
Gill arches, 220
Gish, Duane, 110, 111
Glaessner, M.F., 323
Glossina, 255
Glycolytic enzymes, 169–74, 179, **330;** potential-realizing mutations in amino acid sequences of, 277; structure of, 171–74

Glycoproteins, 257, 258
Goal of evolution, fallacy of, 74–84
Goldschmidt, Richard B., 102, 114–15
Goldschmidt-Clermont, M., 228
Gorgonopsians, 294
Goriki, K., 320
Gould, Stephen Jay, 7, 52, 159, 317, 319
Grafts between cheetahs, skin, 32
Great chain of being, 69, 83, **330**
Great Rift Valley, 85–90, 102
Green, John C., 319
Gryphaea, 75–76
Guanine, 104
Günther, R., 320
Gypsy elements, 233–34, 251

Hadrosaurus, 160
Hagen, F.S., 323
Hahn, Martin, 169–70
Haldane, J.B.S., 34, 319
Hall, Barry G., 7, 260, 324
Halstead, L.B., 324
Hardy, D.E., 14, 318
Harte, P.J., 228
Hawaii, 11–16; Drosophila of, evolution of, 13–16, 288–90; Lepidoptera in, 289
Haynes, J.D., 318
Hazama, R., 320
Heavy (H) chains, 150
Hecht, M.K., 318
Heck, Heinrich, 310–11
Heck, Lutz, 310–11
Hedylepta, 289
Helfand, S.L., 228
Heliconid butterflies, mimicry among, 185–86
Heliconius erato, 185, 192, 197–98
Heliconius melpomene, 185, 192
Helinski, D.R., 317
Helper T cells, 141, 155
Hemoglobin, three-dimensional structure of, 164
Herpes viruses, 155
Heterozygote, 24, **330**
Hiroshima, radiation effects in, 116–19
Hogness, David S., 228, 233, 323

Homeostasis, 5
Homeotic genes, evolution and, 234–38
Homeotic mutants, 229–32, **330**
Hominid fossils, 97–98
Homo erectus, 91, 92, 98, **330**
Homo habilis, 92, **330**
Homologous chromosomes, 146
Homo sapiens, 92
Homozygote, 24, **330**
Hood, Leroy, 317
Hoofed animals, 38
Hooker, Joseph, 71
Hopeful monsters, 115, 217
Horwitz, Marshall S.Z., 244–47, 324
Howell, Stephen H., *xii–xiii,* 317
Hoyle, Fred, 61
Human evolution, discontinuous nature of, 90–93
Human intelligence, evolution of, 314–15
Human longevity, 313–14
Huxley, Julian, 19, 318
Huxley, Thomas Henry, 17
Hybrid dysgenesis, 264–72; P element and, 265–72
Hybridization of DNA, 330
Hypolimnas dinarcha, 198
Hypolimnas dubius, 193–200, **330;** supergene complex in, 197–200, 205

Ice ages, adaptation to, 280–87
IgG, 143
IgM, 143–44
Immune system, 137–49, 247; autoimmune response, 142; cancers of, 148; functioning of, 139–44; generation of final line of defense, 144–49; pressures forcing evolution of, 153–56; response to challenge by trypanosomes, 257–59; secondary immune response, 142–43, 151; structure of, as aid to its evolution, 156–58. *See also* Antibody
Immunoglobulin. *See* Antibody
Inbreeding effect, 118
Individual selection, 331; potential-realizing mutations and, 277;

species selection vs., 50–53, 179, 282–87; speed of, 55–56
Infectious disease, 137–39
Inheritance of acquired characters, 306, 308; Lamarck and, 68–69
Insects, 210–11; environmental change during development of, 231–32; evolutionary split separating vertebrates from, 211, 215–22; exoskeleton and muscle of, 221; fossil record of, 98–99, 222; in Hawaii, evolution of, 13–16, 288–90; mimicry in, 181–209; persistence of segmentation in, 221; resistance to radiation, 120; respiration of, 221; segment duplication in, 235–37; segment evolution in, 237–38; selection acting on evolution of, 219–22; success of, 180–81. *See also specific insects*
Insertion sequence (IS) elements, 124–26, 128, 129; insertion process, 132; random removal from cryptic genes in *E. coli,* 261–62
Intelligence, evolution of human, 314–15
Intrinsic mutational factors, 122–26, 308, **331;** evolution of, 128–31; IS elements, 124–26, 128, 129, 132, 261–62; Mu, 126–30, 133, 252, 268, 332. *See also* Jumping genes
Intron, 172, 177–78, 264, **331;** break up of *D. melanogaster* transposase gene by, 266, 267
IS element, 331. *See also* Insertion sequence (IS) elements

Jardin du Roi (Jardin des Plantes), 65–66
Jaroff, L., 321
Jawbone, evolution of mammalian, 298
Jefferson, Thomas, 60
Jeffery, J., 162
Jenkins, D.G., 80, 320
J gene, 150–53, 156, 157
Johanson, Donald, 86
John Innes Horticultural Institute (England), 251

Jörnvall, Hans, 161–63, 320, 322
Judson, Horace Freeland, 159
Jumping genes, *xi*, 176, 263–64, **331;** in *Antirrhinum majus*, 251–53; bacteriophage Mu, 126–30; environmental shock stimulating, 307, 308; gypsy element, 233–34, 251; IS elements, 124–26, 128, 129, 132, 261–62; McClintock's experiments and unraveling of, 123–24, 249; molecular effect in higher organisms, 131–34; potential-altering mutations and, 278; transposons, 129, 335. *See also* Mobile elements
Jurassic, 80

Kageoka, T., 320
Karch, F., 323
Kaufman, T.C., 303
Keller, E.F., 321
Kilauea (volcano), 11, 12
Killer T cells, 141
Klemenz, R., 323
Koch, Robert, 138
Kovac, R., 320
Kreitman, M.K., 324
Kruif, Paul de, 321

Laboratory organisms, selection for genetic uniformity in, 284–85
Lake Regions of Central Africa, The (Burton), 91
Lake Turkana basin of East Africa, fossil record of freshwater molluscs in, 85–90
Lamarck, Chevalier de (Monet), 65–70, 74, 82
Lamarckism, 15, 68–69, **331**
Lancelet, 220–21
Laski, F.A., 267, 324
Leakey, Louis, 86
Leakey, Mary, 86
Leakey, Richard E., 86, 318
Lepidoptera in Hawaii, 289
Lewin, Roger, 320
Lewis, C.S., 288
Lewis, Edward B., 232–33, 323

Lewis, J., 321
Lewontin, R.C., 325
Ley, Willy, 320
Life: A Search for Order in Diversity, 60–61
Life-span, human, 313–14
Light (L) chains of immunoglobulins, 150
Liljas, A., 174, 322
Limenitis archippus (viceroy), mimicry of monarch butterfly by, 188, 189, 191–93
Linnaean Society, 71
Lipshitz, H.D., 228
Litopterns, 42
Locus, 24, 233, 234, **331**
Loeb, Lawrence A., 244, 324
Lohmann, G.P., 320
London, England, in Victorian Age, 138
London Labour and the London Poor (Mayhew), 138
Longevity, human, 313–14
Lovejoy, A.O., 69, 319
Lyell, Charles, 71
Lymphokines, 141
Lysenkoism, 70

McAuliffe, F.M., 318
McClintock, Barbara, 122–24, 249, 321
McCutchan, T.F., 318
McGinniss, M.H., 318
McKay, Trudy, 269–71, 324
McLaughlin, John C., 294, 295, 325
Macroevolution, 285, **331;** beanbag genetics and, 109; de Vries and, 111–14; Goldschmidt's hopeful monsters and, 115; microevolution and, 108–16
Macrophage, 140, 141, 257, **331**
Madagascar, *Papilio dardanus* found on, 202, 203
Malaria, 25–29, 155
Malarial plasmodia, 257
Malignant tertian malaria, 25–27, 29
Malmgren, Björn A, 93–94, 95, 320
Malthus, T.R., 16
Mammals: of Australia, 37–38, 292; extinction events shaping history

Mammals (*continued*):
of, 281, 291–92; growing diversity in Age of Dinosaurs, 99–100; immune system in, 139–40; major divisions of, 298; Pliocene, 280–83; Pliocene, potential-altering and potential-realizing mutations in, 286–87; of South America, evolution of, 36–44, 290–99

Mammals, Age of, 36, 39, 79, 80, 291

Mani, G.S., 325

Marchalonis, John J., 321

Marriott, Fred, 273–75

Marsupials, 298, 314, **331;** of Australia, 37–38; divergence from placentals, 291; of South America, 39–43

Martin, C., 250, 324

Matjunina, L.V., 224

Maturation rate, selection for slowed human, 313–14

Mauna Kea (volcano), 12

Mauna Loa (volcano), 12

Mayhew, Henry, 138, 321

Maynard, Smith, J. 319

Mayr, Ernst, 284

Medvedev, Zhores A., 319

Mendel, Gregor, 18, 112

Messenger RNA, 148, 223; construction of mature antibody from, 151, 152; IS elements and, 124; promoter sequence and, 244; removal of introns from, 264

Metabolism of therapsids, rate of, 297

Metchnikoff, Elie, 140, 153

Meteorite, extinction of dinosaurs and impact of, 99, 159

Michelson, A.M., 172, 322

Microevolution, 285, **331–32;** macroevolution and, 108–16

Microfossils, 36*n*

Midges, 180

Miller, Louis H., 26, 318

Miller, Oscar, Jr., 20, 21, 318

Miller, Stephan, 324

Mimicry in butterflies, evolution of, 181, 185–209; Batesian, 186–88, 189, 191–93, 208; by *Hypolimnas dubius,* 193–200; of monarch by viceroy, 188, 189, 191–93; Müller-

ian, 184–86, 192; by *Papilio dardanus,* 193, 194, 201–9; supergene complex and, 197–209, 278–79; facilitation of evolution of by supergene complex, 205–9; supergene complex and, in *Hypolimnas dubius* and, 197–200, 205; supergene complex and, in *Papilio dardanus* and, 201–5

Mimics of supergenes, 304–9

Miocene, 80, 280

Mizrokhi, L.J., 224

Mobile elements: gypsy, 233–34, 251; sex and, 263–69; Tam3, 250, 251–53, 334. *See also* Insertion sequence (IS) elements; Jumping genes; Transposable elements

Model species, dependence of Batesian mimic on fate of, 186–88

Molecule, evolutionary facilitation at level of, 161–79

Molluscs, fossil record in Lake Turkana basin of freshwater, 85–90

Monarch butterfly, 188–93; migration of, 189–90; viceroy mimicry of, 188, 189, 191–93

Monet, Jean-Baptiste Antoine de (Lamarck), 65–70, 74, 82

Monotremes, 292, 298

Morata, Ginés, 227, 228, 323

Mouches scramblées, 227–32

Mu, 126–30, 133, **332;** coevolution with host, 129–30; discovery of, 127; effect on *E. coli,* 127, 252; similarity of P element to, 268

Müller, Fritz, 184

Müllerian mimicry, 184–86, 192, **332**

Müllerian ring, 185–86, 192, 197–98

Multicellular organisms, evolution of, 212–14

Multigenic variation, 253, **334**

Murray, J.W., 80, 320

Muscles, insect, 221

Musée d'Histoire Naturelle, 66

Mutagenic agents, 6–7, 179; chemicals as, 121–22; evolution of, 128–31; extrinsic, 116–22; gypsy, 233–34, 251; radiation as, 120–21; somatic mutations in V region and, 157–58; in supergene complex,

200. *See also* Intrinsic mutational factors; Jumping genes
Mutation, 6–7, 102–34, 156–58, **332;** altering model species, 187–88; antigenic switches of trypanosome, 257–59; back, 249; chondrodystrophic dwarf, 104, 113; E (extra legs) mutants, 302, 303; environmental change and, 53, 305–8; in fruit flies, 227–32; in gene pool, 25–30; homeotic, 229–32; intrinsic mutational elements and, 122–26; IS elements and, 124–26, 128, 129, 132, 261–62; jumping genes, molecular effect in other organisms, 131–34; large, idea of, 111–15; microevolution and macroevolution, 108–16, 285, 331–32; Mu and, 126–30, 133, 252, 268, 332; population size and, 33–34; potential-altering, 276–87, 299, 303, 333; potential-realizing, 276–87, 299, 304, 333; somatic, 151, 153, 157–58, 277; somatic, in *Antirrhinum majus,* 249–53; spontaneous, rates of, 130; supergene complex and production of, 199–200
Mycoplasmas, 167
Myoglobin, 297

Nabarro, David, 256
NAD (nicotinamide adenine dinucleotide), 174, 176, **332**
NAD-binding enzyme, 176–77
NADH, 170, 174
Nagana, 255–57
Nagasaki, radiation effects in, 116–19
National Institutes of Health, 26, 32
Natural selection, 16–19, 70; Darwin and, 16–19, 71, 73; large mutations and, 113; power of, 107; standard answer to creationist argument using, 243; for status quo, 77
Nature (journal), 72
Neanderthal man, 332
Neel, J.V., 320
Nematodes, 219

Neo-Darwinism, 19–25, 44
Neurospora, 114
Newtonian physics, 70
Notochord, 218
Nuclear parasites, 133
Nucleotides, 22–23
Nymphalidae, 189; *Hypolimnas dubius,* 193–200; viceroy butterfly, 188, 189, 191–93

O'Brien, Steven J., 32, 318
Opossum, 37
Orchids, mimicry of female of pollinating insect species, 183
Ordovician, 78
Oregon-R strain of *Drosophila melanogaster,* 300
Orgel, L.E., 321
Origin of Species, The (Darwin), 16, 70, 72n
Orkin, S.H., 172, 322
Orthogenesis, 74–76, **332**
Otake, M., 320
Overbaugh, Julie, 324
Overton, William, 61–62
Ow, D.W., 317
Owen, D.F., 194, 322

Pair-rule genes, 237
Pangaea, 332; I, 211, 212, 216; II, 211–12
Papilio dardanus, 193, 194, 201–9, **332**
Papilionidae, 201
Paquin, Charlotte E., 132, 321
Paralichthys albigattus, camouflage ability of, 182
Parallel evolution, 41, 290–91, **332–33;** evolutionary facilitation and, 291, 298–99
Parasites: benign vs. malignant malaria and, 25–29; evolution of immune system to defend against, 154–56; in human DNA, 7, 133–34. *See also* Immune system
Parasitic DNA, 126
Pardue, Mary Lou, 226, 323
Parrington, D.R., 325

Parthenogenesis, 54–55
Pasteur, Louis, 138, 169, 170
Pauling, Linus, 171
P element, 234, 265–72, 308, **332;** confinement to germ line, 266–68; selection experiments on, 269–72
Pelycosaurs, 292, 293, 314, **333**
Permian, 292, 294
Permian-Triassic transition, 294–96
Persson, B., 162
Perutz, Max, 159
Pfeifer, M., 323
Phagocytes, 140, 153
Phenotype, 23
Philosophie zoologique (Lamarck), 66
Phosphoglycerate kinase, 172
Photosynthetic bacteria, 212
Picture-winged flies, 14
Placentals, 37–38, 40–42, 291, 298, 314. *See also* Mammals
Planktonic species, 78
Plasmodia, malarial, 257
Platypus, duckbill, 292
Pleistocene, 333; ice age, adaptations to, 280–87
Pliocene, 280–83, **333;** Foraminifera from, 80; mammals of, 280–83; mammals of, potential-altering and potential-realizing mutations in, 286–87
Pliohippus, 42
Pneumococcus, 154
Pneumonia, 154
Polymorphisms, genetic, 206
Polysaccharide coat on pneumococcus, 154
Popko, B., 317
Population genetics, 34–35, 285
Population size, mutations and, 33–34
Potassium-40, 86
Potential-altering mutations, 276–87, 299, 303, **333**
Potential-realizing mutations, 276–87, 299, 304, **333**
Precambrian, 215–18, 247–48, **333**
Predators, evolution of mimicry affecting evolution of, 181–83
Primitive soup, 168
Proboscipedia (*D. melanogaster* mutant), 231

Promoter sequence, 244–47, **333;** disruption of, by Tam3 in *Antirrhinum,* 251–53
Prophage, 127
Protein domains, 179
Proteins: conservation of three-dimensional structure in course of evolution of, 163–64; genes for, 23; types of structure, 171. *See also* Amino acid; Antibody
Punctuated equilibrium, 52–53, **333–34**
Purgatorius, 37
Pyrotheria, 42–43

Quantitative (multigenic) variation, 253, **334**
Quest for Fire (film), 91

Racial group, genetic diversity within, 286
Radiation, adaptive. *See* Adaptive radiation
Radiation, mutational changes caused by, 116–21
Radiation Effects Research Foundation, 116
Radiolaria, 93
Raff, M., 321
Raff, R.A., 303
Random DNA sequences, experiment with, 244–48
Recombination, genetic, 34, 114, 146, 147
Redhead, C., 317
Rehnquist, William, 63, 64
Religion, evolution and, 9–10, 59–65. *See also* Creationism
Reptiles, 294
Reptiles, Age of, 212, 291
Restriction enzymes, 300
Retrovirus, 133, 233, **334;** gypsy element, 233–34, 251
Rio, D.C., 267, 324
RNA, 20–22, 177, **334;** electron micrograph of DNA strand making, 21; messenger RNA, 124, 148, 151, 152, 223, 244, 264

Roberts, K., 321
Rossmann, Michael G., 173, 174, 322
Rubin, Gerald M., 263, 266, 267, 324

Saavedra, R.A., 317
Sabertooths, 44; placental and marsupial, 40–42
Saint, R.B., 228
Salicin, 260–62
Salicylic acid, 260
Sánchez-Herrero, E., 323
Sapienza, C., 321
Satoh, C., 320
Scalia, A., 63, 64
Schistosome worms and schistosomiasis, 155, 257
Schliemann, Heinrich, 160
Schneuwly, Stephan, 231, 323
Schopf, T.J.M., 319
Schull, W.J., 320
Science, creationism as, 59–65
Scopes trial (1925), 62
Sea pens, 213
Secondary immune response, 142–43, 151
Segmentation: in chordates, 220–21; evolution of in insects, 237–38; persistence in insects vs. vertebrates, 221
Segment duplication, 235–37
Segmented worms, 219–20
Seilacher, A., 323
Selection, 83; natural, 16–19, 70, 71, 73, 77, 107, 113, 243; somatic, 151, 153. *See also* Individual selection; Species selection
Selective breeding, 310–13
Semmelweis, Ignaz, 138–39
Serpollet, Léon, 274
Sexual reproduction: evolution of, 56; genes passed on by, 24; mobile element and, 263–69; species selection and, 53–55
Shapiro, James A., 321
Shaw, George Bernard, 46
Sheppard, Philip, 203–5, 322
Shine, H.D., 317
Shirioshi, T., 318

Sidman, R.L., 317
Silkworm moth, 301–2, 303, 304
Simpson, George Gaylord, 39, 41, 76, 319
Singer, M.C., 322
Siphonophores, 213–14
Skin grafts between cheetahs, 32
Sleeping sickness, African, 255–57
Sloths, 38–39, 42
Slow viruses, 155
Small nuclear ribonucleoproteins (snurps), 264
Smilodon, 40, 41
Snapdragon, 301; *Antirrhinum majus*, somatic mutations in, 249–53; evolutionary toolbox of, 248–54
Snurps, 264
Social Darwinism, 17
Somatic mutation, 277, **334;** in *Antirrhinum majus*, 249–53; production of antibodies and, 151, 153, 157–58
Somatic selection, 151, 153
South America, evolution of mammals in, 36–44, 290–99
Speciation, process of, 84–96
Species: defined, 51; endemic, 329; environmental change and evolution of, 52–53, 84–96; measuring DNA variation within, 285, 286; short-lived, 89; variation in internal organization within, 209
Species selection, 334; individual selection vs., 50–53, 179, 282–87; sexual reproduction and, 53–55; slowness of, 55, 56, 57
Spencer, Herbert, 17
Spierer, P., 323
Spieth, H.T., 14, 318
Spiny anteater, 292
Spontaneous generation, 67
Spontaneous mutation, rates of, 130
Stanley, F.E. and F.O., 273–75
Stanley, Steven M., 52, 53, 319
Stanley Steamer automobile, 273–75
Starling, E.H., 4
Steere, W.C., 318
Stone, W.S., 14, 318
Stress, body's means of coping with, 4–5
Stromatolites, 214

Struhl, Gary, 236, 323
Sub-Saharan Africa, *Papilio dardanus*
found in, 201–5
Sunset Boulevard (film), 167
Supergene complex, 227–31, 276–
79, **334**; in *Hypolimnas dubius,*
197–200, 205; mimicry, evolutionary facilitation by, 205–9; genes
mimicking, 304–9; in *Papilio dardanus,* 201–5. *See also* Antennapedia complex (ANT-C); Bithorax
complex (BX-C)
Survival of the fittest, 17
Susman, R.L., 320

Takahashi, N., 317, 320
Tam3, 250, 251–53, **334**
Tanaka, Y., 325
Tasmanian wolf, 40, 41
Taxonomy, science of, 66
Taylor, Austin L., 127, 321
T cell, 140–41, 155, **334;** helper,
141, 155; killer, 141; receptors,
140, 141
Teeth, therapsid, 294–97
Teleology, 334
Temperature, activity of jumping
genes and, 132
"Temporary" species, 89
Tetracycline, experiment on gene
conferring resistance of *Esche-
richia coli* to, 244–47
Thalidomide babies, 232
Therapsids, 37, 291–98, 314, **334;**
adaptive radiations of, 294–97;
metabolic rate of, 297; potentially
mammal-like, 297
Thompson, Silvanus P., 135
Thrinaxodon, 295, 297
Thylacosmilus, 40–42
Thymine, 104
Thymus, 140
Tn10 (transposon), 129
Toolboxes. *See* Evolutionary tool-
boxes
Topaloglou, A., 320
Transposable element, 158, 268,
335; discovery of, 130–31; envi-
ronmental shock triggering

movement of, 308; removal of
from genome, 250, 251–53. *See
also* Jumping genes; Mobile ele-
ments
Transposase, 265–67
Transposon, 129, **335**
Transposon Antirrhinum majus
(Tam3) element, 250, 251–53, 334
Triassic, 298, **335;** transition to,
294–96
Tribolium, 302–3
Trypanosome, 254–59; antigenic
switching by, 257–59; coevolution
of host with, 256–57
Tsetse flies, 255–56, 258, 259
Tuna, metabolism of, 297
Tunicates, 218–19
Turner, J.R.G., 322
Tuschl, H., 320

Unequal crossing-over, 146, 147,
176
Ungulates, 38
Uniformity, genetic, in species,
283–87
U.S. Department of Agriculture,
302
U.S. Supreme Court, on equal-time
law, 62, 63
Urus (aurochs), 310–12

Valentine, James W., 216–17, 323
Variability. *See* Genetic variability
Variable (V) region of immunoglobulin molecule, 150, 151, 153, 157–
58
Varves, 215
Velella, 213
Vertebrates, 218; environmental
shock during development of,
232; evolutionary split from insects, 211, 215–22; selective pressures leading to evolution of,
217–19
V gene, 150, 152, 153, 156–58, 277
Viceroy butterfly, mimicry of monarch by, 188, 189, 191–93

Victorian Age, London in, 138
Viruses: AIDS, 155–56, 254; bacteriophage Mu, 126–30, 133, 252, 268, 332; common cold, 155; computer, 125–26; evolution of immune system to handle, 154–56; herpes, 155; slow, 155
Vitamin C, loss of gene for manufacture of, 30
Vivax malaria, 25–29
Volcanic ash, dating of, 86
Vrba, Elizabeth, 317
V region, 150, 151, 153, 157–58

Waddington, C.H., 305–8, 325
Walking down the chromosome, 233
Wallace, Alfred Russel, 70–73, 186, 319
Wallace, Bruce, 120, 320
Watch, evolution of, 165
Watson, J.D., 321
Weinberg, Steven, 323
White, D.O., 321
White blood cells, 140, 141

White Cliffs of Dover, 79
Whitehead, Alfred North, 45
Wickler, W., 322
Wickramasinghe, Chandra, 61
Wigglesworth, V.B., 210
Wildlife Safari Park (Oregon), 31
Wildt, D.E., 318
Wild type, 305; of laboratory organisms, 284
Wilford, J.N., 322
Williamson, Peter G., 85–90
Williamson, Valerie M., 132, 321
Wills, C., 320, 325
Wisdom of the Body, The (Cannon), 4
Wood, K.V., 317
Worms: schistosome, 155, 257; segmented, 219–20
Wright, Sewall, 34

X-ray crystallography, 164, **335**

Zymase, 170. *See also* Glycolytic enzymes